Soil and Applied Microbiology

The Author

Dr. V. Ponvizhi Ramya has completed her graduation from Tamil Nadu Agricultural University in 2009. She has done her *M.Sc.* and *Ph.D* programme in Agricultural Microbiology from the same University. She has received a National Award from ICAR and NAIP for her innovative idea in her Doctoral research. She has published many research articles in reputed journals. Currently she is working as Assistant Professor in Agricultural Microbiology at Thanthai Roever Institute of Agriculture and Rural Development, Chennai - Trichy Hwy, Tamil Nadu.

Soil and Applied Microbiology

V. Ponvizhi Ramya

2020

Associated Publishing Company®

A Division of

Astral International Pvt. Ltd.

New Delhi – 110 002

ISBN 9789390371730 (Int. Edition)

Published by : **Associated Publishing Company®**
A Division of
Astral International Pvt. Ltd.
– ISO 9001:2015 Certified Company –
4736/23, Ansari Road, Darya Ganj
New Delhi-110 002
Ph. 011-43549197, 23278134
E-mail: info@astralint.com
Website: www.astralint.com

Preface

It gives great pleasure and satisfaction to fulfill the long felt need of microbiologists and students for a comprehensive on Soil and Applied Microbiology. Experience gained from teaching agricultural microbiology encouragement gained from the university, faculty members and library facilities have made the dream come true. The modern metagenomic approach, PGPR and PPFM have been present in a readable form so that it will be student friendly. This book will definitely pave the way for the young aspiring microbiologist mind to take up the subject with confidence.

Several of my colleagues have contributed in bringing out this book. My family members has helped in preparing this script and I greatly indebted to all of them.

Dr. V. Ponvizhi Ramya

Contents

Unit - IV: Microbial Biotechnology

Unit - I: Soil Microbiology

1

Introduction to Soil Microbiology

Introduction

Soil represents a medium or substrate in which numerous microorganisms live and bring about a great variety of processes that are responsible for continuation of the cycle of life in nature. The numerous living forms which spend their life in soil, ranging from sub microscopic forms to the lower animal forms. With the growing recognition of the numerous processes carried out by the microorganisms in the soil there gradually emerged a branch of microbiology, which came to known as soil microbiology. It is a branch of soil science concerned with soil inhabiting microorganisms and their functions and activities.

Since soil microbiology concerns with soil microorganisms and their processes it is closely associated with soil biochemistry. Medical bacteriologists were interested in the soil as a medium for the growth and survival of disease producing organisms. Agricultural chemists are also interested in the soil processes that result from the activities of microorganisms. General bacteriologist, zoologist, botanist were interested in certain special group of organisms found in soil. Recently, soil microbiology has expanded to include the study of the role of soil microorganisms in genetic engineering, in the biological control of pests and diseases, the degradation of pollutants, production and destruction of radioactive gases and its transfer. Thus microbial participation in several important processes emphasizes that soil microbiology has become a global science.

Soil Microbiology

It is branch of science/microbiology which deals with study of soil microorganisms and their activities in the soil.

- Deals with the microscopic organisms of the soil
- Their population and activities
- Role of microorganisms in various transformations taking place in the soil and
- Their importance in plant nutrition and crop production

Soil

It is the outer, loose material of earth's surface which is distinctly different from the underlying bedrock and the region which support plant life. Agriculturally, soil is the region which supports the plant life by providing mechanical support and nutrients required for growth. From the microbiologist view point, soil is one of the most dynamic sites of biological interactions in the nature. It is the region where most of the physical, biological and biochemical reactions related to decomposition of organic weathering of parent rock take place.

Components of Soil

Soil is an admixture of five major components *viz.* organic mater, mineral matter, soil-air, soil water and soil microorganisms/living organisms. The amount/proposition of these components varies with locality and climate.

1. **Mineral / Inorganic Matter:** It is derived from parent rocks/bed rocks through decomposition, disintegration and weathering process. Different types of inorganic compounds containing various minerals are present in soil. Amongst them the dominant minerals are Silicon, Aluminum and iron and others like Carbon, Calcium Potassium, Manganese, Sodium, Sulphur, Phosphorus *etc.* are in trace amount. The proportion of mineral matter in soil is slightly less than half of the total volume of the soil.
2. **Organic matter/components:** Derived from organic residues of plants and animals added in the soil. Organic matter serves not only as a source of food for microorganisms but also supplies energy for the vital processes of metabolism which are characteristics of all living organisms. Organic matter in the soil is the potential source of N, P and S for plant growth. Microbial decomposition of organic matter releases the unavailable nutrients in available from. The proportion of organic matter in the soil ranges from 3-6% of the total volume of soil.
3. **Soil Water:** The amount of water present in soil varies considerably. Soil water comes from rain, snow, dew or irrigation. Soil water serves as a solvent and carrier of nutrients for the plant growth. The microorganisms inhabiting in the soil also require water for their metabolic activities. Soil water thus, indirectly affects plant growth through its effects on soil and microorganisms. Percentage of soil-water is 25% total volume of soil.
4. **Soil air (Soil gases):** A part of the soil volume which is not occupied by soil particles *i.e.* pore spaces are filled partly with soil water and partly with soil air. These two components (water & air) together only accounts for approximately half the soil›s volume. Compared with atmospheric air, soil is lower in oxygen and higher in carbon dioxide, because CO_2 is continuous recycled by the microorganisms during the process of decomposition of organic matter. Soil air comes from external atmosphere and contains nitrogen, oxygen CO_2 and water vapour (CO_2> oxygen). CO_2 in soil air (0.3-1.0%) is more than atmospheric air (0.03%). Soil aeration plays important role in plant growth, microbial population, and microbial activities in the soil.

5. **Soil microorganisms:** Soil is an excellent culture media for the growth and development of various microorganisms. Soil is not an inert static material but a medium pulsating with life. Soil is now believed to be dynamic or living system.

Soil contains several distinct groups of microorganisms and amongst them bacteria, fungi, actinomycetes, algae, protozoa and viruses are the most important. But bacteria are more numerous than any other kinds of microorganisms. Microorganisms form a very small fraction of the soil mass and occupy a volume of less than one percent. In the upper layer of soil (top soil up to 10-30 cm depth *i.e.* Horizon A), the microbial population is very high which decreases with depth of soil. Each organisms or a group of organisms are responsible for a specific change / transformation in the soil. The final effect of various activities of microorganisms in the soil is to make the soil fit for the growth & development of higher plants. Living organisms present in the soil are grouped into two categories as follows.

1. Soil flora (micro flora) *e.g.* Bacteria, fungi, Actinomycetes, Algae and
2. Soil fauna (micro fauna) animal like *e.g.* Protozoa, Nematodes, earthworms, moles, ants, rodents.

Relative proportion / percentage of various soil microorganisms are: Bacteria-aerobic (70%), anaerobic (13 %), Actinomycetes (13%), Fungi /molds (03 %) and others (Algae Protozoa viruses) 0.2-0.8 %. Soil organisms play key role in the nutrient transformations.

Soil Humus

Humus is the organic residue in the soil resulting from decomposition of plant and animal residues in soil, or it is the highly complex organic residual matter in soil which is not readily degraded by microorganism, or it is the soft brown/dark coloured amorphous substance composed of residual organic matter along with dead microorganisms.

Composition of Humus

In most soil, percentage of humus ranges from 2-10 percent, whereas it is up to 90 percent in peat bog. On average humus is composed of Carbon (58 %), Nitrogen (3-6 %, Av.5%), acids - humic acid, fulvic acid, humin, apocrenic acid, and C: N ratio 10:1 to 12:1. During the course of their activities, the microorganisms synthesize number of compounds which plays important role in humus formation.

Functions/Role of Humus

1. It improves physical condition of soil
2. Improve water holding capacity of soil
3. Serve as store house for essential plant nutrients
4. Plays important role in determining fertility level of soil
5. It tend to make soils more granular with better aggregation of soil particles
6. Prevent leaching losses of water soluble plant nutrients

7. Improve microbial/biological activity in soil and encourage better development of plant-root system in soil
8. Act as buffering agent *i.e.* prevent sudden change in soil PH/soil reaction
9. Serve as source of energy and food for the development of soil organisms
10. It supplies both basic and acidic nutrients for the growth and development of higher plants
11. Improves aeration and drainage by making the soil more porous

Distinct Phases of Soil Microbiology

1. Ecological phase

Study on the quantitative and qualitative composition of the microscopic and ultramicroscopic soil population.

2. Experimental or physiological phase

Study on the physiology and biochemistry of the organisms, their role in the cycle of life in nature and their utilization for the formation of valuable metabolic products.

3. Agronomical phase

Application of microbiological activities for the enhancement of soil fertility and crop production.

4. Pedological phase

Importance of microorganisms in soil formation and soil structure.

Scope and Importance of Soil Microbiology

Living organisms both plant and animal types constitute an important component of soil. Though these organisms form only a fraction (less than one percent) of the total soil mass, but they play important role in supporting plant communities on the earth surface. While studying the scope and importance of soil microbiology, soil-plant-animal ecosystem as such must be taken into account. Therefore, the scope and importance of soil microbiology, can be understood in better way by studying aspects like

1. Soil as a living system
2. Soil microbes and plant growth
3. Soil microorganisms and soil structure
4. Organic matter decomposition
5. Humus formation
6. Biogeochemical cycling of elements
7. Soil microorganisms as bio-control agents
8. Soil microbes and seed germination

9. Biological N_2 fixation
10. Degradation of pesticides in soil.

1. **Soil as a living system:** Soil inhabit diverse group of living organisms, both micro flora (fungi, bacteria, algae and actinomycetes) and micro-fauna (protozoa, nematodes, earthworms, moles, ants). The density of living organisms in soil is very high *i.e.* as much as billions / gm of soil, usually density of organisms is less in cultivated soil than uncultivated / virgin land and population decreases with soil acidity. Top soil, the surface layer contains greater number of microorganisms because it is well supplied with Oxygen and nutrients. Lower layer / subsoil is depleted with Oxygen and nutrients hence it contains fewer organisms. Soil ecosystem comprises of organisms which are both, autotrophs (Algae, BOA) and heterotrophs (fungi, bacteria). Autotrophs use inorganic carbon from CO_2 and are "primary producers" of organic matter, whereas heterotrophs use organic carbon and are decomposers/consumers.
2. **Soil microbes and plant growth:** Microorganisms being minute and microscopic, they are universally present in soil, water and air. Besides supporting the growth of various biological systems, soil and soil microbes serve as a best medium for plant growth. Soil fauna & flora convert complex organic nutrients into simpler inorganic forms which are readily absorbed by the plant for growth. Further, they produce variety of substances like IAA, gibberellins, antibiotics *etc.* which directly or indirectly promote the plant growth
3. **Soil microbes and soil structure:** Soil structure is dependent on stable aggregates of soil particles-Soil organisms play important role in soil aggregation. Constituents of soil are *viz.* organic matter, polysaccharides, lignins and gums, synthesized by soil microbes plays important role in cementing / binding of soil particles. Further, cells and mycelial strands of fungi and actinomycetes, Vormicasts from earthworm is also found to play important role in soil aggregation.

   Different soil microorganisms, having soil aggregation / soil binding properties are graded in the order as Fungi > Actinomycetes > Gum producing Bacteria > Yeasts. Examples are: Fungi like Rhizopus, Mucor, Chaetomium, Fusarium, Cladasporium, Rhizoctonia, Aspergillus, Trichoderma and Bacteria like Azofobacler, Rhizobium Bacillus and Xanlhomonas.
4. **Soil microbes and organic matter decomposition:** The organic matter serves not only as a source of food for microorganisms but also supplies energy for the vital processes of metabolism that are characteristics of living beings. Microorganisms such as fungi, actinomycetes, bacteria, protozoa *etc.* and macro organisms such as earthworms, termites, insects *etc.* plays important role in the process of decomposition of organic matter and release of plant nutrients in soil. Thus, organic matter added to the soil is converted by oxidative decomposition to simpler nutrients / substances for plant growth and the residue is transformed into humus. Organic matter / substances include cellulose, lignins and proteins (in cell wall

of plants), glycogen (animal tissues), proteins and fats (plants, animals). Cellulose is degraded by bacteria, especially those of genus Cytophaga and other genera (Bacillus, Pseudomonas, Cellulomonas, and Vibrio Achromobacter) and fungal genera (Aspergillus, Penicilliun, Trichoderma, Chactomium, Curvularia). Lignins and proteins are partially digested by fungi, protozoa and nematodes. Proteins are degraded to individual amino acids mainly by fungi, actinomycetes and Clostridium. Under unaerobic conditions of waterlogged soils, methane are main carbon containing product which is produced by the bacterial genera (strict anaerobes) Methanococcus, Methanobacterium and Methanosardna.

5. **Soil microbes and humus formation:** Humus is the organic residue in the soil resulting from decomposition of plant and animal residues in soil, or it is the highly complex organic residual matter in soil which is not readily degraded by microorganism, or it is the soft brown/dark coloured amorphous substance composed of residual organic matter along with dead microorganisms.

6. **Soil microbes and cycling of elements:** Life on earth is dependent on cycling of elements from their organic / elemental state to inorganic compounds, then to organic compounds and back to their elemental states. The biogeochemical process through which organic compounds are broken down to inorganic compounds or their constituent elements is known "Mineralization", or microbial conversion of complex organic compounds into simple inorganic compounds & their constituent elements is known as mineralization.

   Soil microbes plays important role in the biochemical cycling of elements in the biosphere where the essential elements (C, P, S, N & Iron *etc.*) undergo chemical transformations. Through the process of mineralization organic carbon, nitrogen, phosphorus, Sulphur, Iron *etc.* are made available for reuse by plants.

7. **Soil microbes and biological N_2 fixation:** Conversion of atmospheric nitrogen in to ammonia and nitrate by microorganisms is known as biological nitrogen fixation.

Fixation of atmospheric nitrogen is essential because of the reasons:

1. Fixed nitrogen is lost through the process of nitrogen cycle through denitrification.
2. Demand for fixed nitrogen by the biosphere always exceeds its availability.
3. The amount of nitrogen fixed chemically and lightning process is very less (*i.e.* 0.5%) as compared to biologically fixed nitrogen
4. Nitrogenous fertilizers contribute only 25% of the total world requirement while biological nitrogen fixation contributes about 60% of the earth's fixed nitrogen
5. Manufacture of nitrogenous fertilizers by "Haber" process is costly and time consuming.

The numbers of soil microorganisms carry out the process of biological nitrogen fixation at normal atmospheric pressure (1 atmosphere) and temperature (around 20 °C).

Two groups of microorganisms are involved in the process of Biological Nitrogen Fixation.

A. Non-symbiotic (free living)

B. Symbiotic (Associative)

A. Non-symbiotic (free living)

Depending upon the presence or absence of oxygen, non symbiotic N_2 fixation prokaryotic organisms may be aerobic heterotrophs (Azotobacter, Pseudomonas, Achromobacter) or aerobic autotrophs (Nostoc, Anabena, Calothrix, BGA) and anaerobic heterotrophs (Clostridium, Kelbsiella. Desulfovibrio) or anaerobic Autotrophs (Chlorobium, Chromnatium, Rhodospirillum, Meihanobacterium *etc*)

B. Symbiotic (Associative)

The organisms involved are Rhizobium, Bratfyrhizobium in legumes (aerobic): Azospirillum (grasses), Actinonycetes Frankia(with Casuarinas, Alder).

Soil microbes as biocontrol agents: Several ecofriendly bioformulations of microbial origin are used in agriculture for the effective management of plant diseases, insect pests, weeds *etc.* eg: Trichoderma sp and Gleocladium sp are used for biological control of seed and soil borne diseases. Fungal genera Entomophthora, Beauveria, Metarrhizium and protozoa Maltesia grandis. Malameba locustiae *etc* are used in the management of insect pests. Nuclear polyhydrosis virus (NPV) is used for the control of Heliothis / American boll worm. Bacteria like Bacillus thuringiensis, Pseudomonas are used in cotton against Angular leaf spot and boll worms.

Degradation of pesticides in soil by microorganisms: Soil receives different toxic chemicals in various forms and causes adverse effects on beneficial soil micro flora / micro fauna, plants, animals and human beings. Various microbes present in soil act as the scavengers of these harmful chemicals in soil. The pesticides/chemicals reaching the soil are acted upon by several physical, chemical and biological forces exerted by microbes in the soil and they are degraded into non-toxic substances and thereby minimize the damage caused by the pesticides to the ecosystem. For example, bacterial genera like Pseudomonas, Clostridium, Bacillus, Thiobacillus, Achromobacter *etc.* and fungal genera like Trichoderma, Penicillium, Aspergillus, Rhizopus, and Fusarium are playing important role in the degradation of the toxic chemicals / pesticides in soil.

Biodegradation of hydrocarbons: Natural hydrocarbons in soil like waxes, paraffin's, oils *etc* are degraded by fungi, bacteria and actinomycetes. *E.g.* ethane (C_2H_6) a paraffin hydrocarbon is metabolized and degraded by Mycobacteria, Nocardia, Streptomyces Pseudomonas, Flavobacterium and several fungi.

2

History and Development of Soil Microbiology

History of Soil Microbiology (1600 - 1920)

There is enough evidence in the literature to believe that microorganisms were the earliest of the living things that existed on this planet. Man depends on crop plants for his existence and crop plants in turn depend on soil and soil microorganisms for their nutrition. Scientists form the beginning studied the microorganisms from water, air, soil *etc.* and recognized the role of microorganisms in natural processes and realized the importance of soil microorganisms in growth and development of plants.

Thus, we see that microorganisms have been playing a significant role long before they were discovered by man. Today, soil is considered to be the main source of scavenging the organic wastes through microbial action and is also a rich store house for industrial micro flora of great economic importance.

Unlike soil science whose origin can be traced back to Roman & Aryan times, soil microbiology is emerged as a distinct branch of soil science during first half of the 19th century. Some of the notable contributions made by several scientists in field of soil microbiology are highlighted in the following paragraphs.

A. V. Leeuwenhock (1673) discovered and described microorganisms through his own made first simple microscope with magnification of 200 to 300 times. He observed minute, moving objects which he called **"animalcules"** (small animals) which are now known as protozoa, fungi and bacteria. He for the first time made the authentic drawings of microorganisms (protozoa, bacteria, fungi).

Robert Hook (1635-1703) developed a compound microscope with multiple lenses and described the fascinating world of the microbes.

J. B. Boussingault (1838) showed that leguminous plants can fix atmospheric nitrogen and increase nitrogen content in the soil.

J. Von Liebig (1856) showed that nitrates were formed in soil due to addition of nitrogenous fertilizers in soil.

S. N. Winogradsky discovered the autotrophic mode of life among bacteria and established the microbiological transformation of nitrogen and sulphur. Isolated for the first time nitrifying bacteria and demonstrated role of these bacteria in nitrification (1890), further he demonstrated that free-living *Clostridium pasteuriamum* could fix atmospheric nitrogen (1893). Therefore, he is considered as **"Father of Soil Microbiology"**.

W. B. Leismaan (1858) and M. S. Woronin (1866) demonstrated that root nodules in legumes were formed by a specific group of bacteria.

Jodin (1862, France) gave the first experimental evidence of elemental nitrogen fixation by microorganisms.

Robert Koch (1882) developed gelatin plate/ streak plate technique for isolation of specific type of bacteria in soil, formulated Koch›s postulates to establish causal relationship between host - pathogen and disease.

R. Warington (1878) showed that nitrification in soil was a microbial process.

B. Frank i) discovered (1880) an actinomycetes "Frankia" (Actinorhizal symbiosis) inducing root nodules in non-legumes tress of genera *Alnus sp* and *Casurina* growing in temperate forests, ii) coined (1885) the term « Mycorrhiza» to denote association of certain fungal symbionts with plant roots (Mycorrhiza-A symbiotic association between a fungus and roots of higher plants. Renamed the genus Bacillus as Rhizobium (1889).

H. Hellriegel and H. Wilfarth (1886) showed that the growth of non-legume plant was directly proportional to the amount of nitrogen supplied, whereas, in legumes there was no relationship between the quantity of nitrogen supplied and extent of plant growth. They also suggested that bacteria in the root nodules of legumes accumulate atmospheric nitrogen and made it available to plants. Showed that a mutually beneficial association exists between bacteria (*Rhizobia*) and legume root and legumes could utilize atmospheric nitrogen (1988).

Beijerinck, Martinus Willem (1851-1931)

M. W. Beijerinck (1888) isolated root nodule bacteria in pure culture from nodules in legumes and named them as *Bacillus radicola* Considered as **father of "Microbial ecology"**. He was the first Director of the Delft School of microbiology (Netherland).

Beijerinck was born in Amsterdam and studied chemical engineering at the Delft Polytechnic, where, after working as a bacteriologist with an industrial company, he taught and carried out research from 1895.He published his finding in 1898 that an agent smaller than bacteria could cause diseases, an agent that he called a virus. In the 1880s and 1890s Beijerinck studied the disease that stunts the growth of tobacco plants and mottles their leaves (now called the tobacco mosaic virus disease). Trying to isolate a causative agent, he pressed out the juice of infected tobacco leaves and found that the juice alone was able to infect healthy plants, even after he had passed it through a filter that removed bacteria. Beijerinck was also certain that the causative agent was not a toxin because he could infect a healthy

plant and from that plant infect another healthy plant, continuing this process indefinitely - therefore the infective agent had to be capable of reproduction. He believed that the filtered juice of the infected plants was itself alive, and he called the causative agent a filterable virus (the Latin word for poison). First isolated several soil bacteria *viz.,Azotobacter chroococcum, Thiobacillus thioparus.* Isolated first *Azospirillum* from *Paspalum* roots in Brazil and named it as *Azotobacter paspali* in 1922. He developed **enrichment culture** technique, a way to isolate microbes with certain growth preferences. Using enrichment media, it was possible to isolate enormous variety of bacteria with different physiological capabilities and different growth ranges. This demonstrated the impressive variety of microbial lifestyles, much broader than that of animals or plants.

Beijerinck and Winogradsky (1890) developed the enrichment culture technique for isolation of soil organisms, proved independently that transformation of nitrogen in nature is largely due to the activities of various groups of soil microorganisms (1891). Therefore, they are considered as **«Pioneer›s in soil bacteriology".**

Sergei Winogradsky (1856-1953)

Winogradsky extended awareness of microbial diversity. He discovered bacteria capable of **autotrophic** ("self-feeding") growth using inorganic compounds such as H_2S as their only energy source, and CO_2 as their only C-source. This was the first discovery that organisms other than green plants or algae could exist without consuming organic matter. Also discovered groups of photosynthetic bacteria that do not produce oxygen as their waste product. Developed culture technique known as **Winogradsky column**, in which mud and water are left in glass tube exposed to light. Over time, different microbial communities grow and interchange waste products and nutrients. He Established N & S transformation in soil.1890 and demonstrated the role of bacteria in nitrification.**S. N. Winogadsky (1891)** demonstrated the role of bacteria in nitrification and further in fill 1983 demonstrated that free living *Clostridium pasteurianum* could fix atmospheric nitrogen.

Omeliansky (1902) found the anaerobic degradation of cellulose by soil bacteria.

J. G. Lipman and P. E. Brown (1903, USA) studied ammonification of organic nitrogenous substances by soil microorganisms and developed the **Tumbler or Beaker** for studying different types of transformation in soil.

Hiltner (Germany, 1904) coined the term **"Rhizosphere"** to denote that region of soil which is subjected to the influence of plant roots. Rhizosphere is the region where soil and plant roots make contact.

Russel and Hutchinson (1909, England), proved the importance of protozoa controlling/ maintaining bacterial population and their activity in soil.

Conn (1918) developed **"Direct soil examination"** technique for studying soil microorganisms.

Rayner (192I) and Melin (1927) carried out the intensive study on Mycorrhiza.

The fertility of soil depends not only on its chemical composition but also on the qualitative and quantitative nature of microorganisms inhabiting it. Strictly speaking, the development of microbiology as a branch of science can be dated back to the time of people who ground lens from glass and saw microorganisms, through them.

Anton Van Leeuwenhoek, a linen draper from Holland (1632-1723) is credited with having made the first authenticated drawings of microorganisms which had its roots in the age of Aristotle (384-322 B.C) and which was revoked by the so-called experimental evidence of John Needham (1713-1781) was scattered by the conclusive experimental findings of Louis Pasteur (1822-1895). This was followed by Robert Koch's Koch postulates (1843-1910) and the pure culture technique of Joseph Lister (1878).

Soil microbiology emerged a distinct branch of soil science only in 1838 after the French agricultural chemist and farmer, J.B. Boussingault showed that legumes can obtain nitrogen from air when grown in soil which was not heated. Since then several significant developments made in soil microbiology are in the following.

Selman.A.Waksman (1944)

In 1922, Waksman isolated *Thiobacillus thioxidans.* He discovered the antibiotic Streptomycin in the year 1944 for tuberculosis - *Mycobacterium tuberculosis.* Also discovered Neomycin, Actinomycin antibiotics. He wrote a book on Principles of Soil Microbiology. In 1942, he showed the importance of soil as the source of antagonistic organisms. He also studied variety of biochemical reactions carried out by soil microorganisms while decomposing organic matter.

S. A. Waksman published the book **"Principles of soil Microbiology"** and thereby encouraged the research in soil microbiology (1927). Studied the role of soil as the source of antagonistic organisms with special reference to soil actinomycetes (1942) and discovered the antibiotic «Streptomycin» produced by *Streptomyces griseus*, a soil actinomycets (1944).

Rossi (1929) and Cholondy (1930) developed **"Contact Slide / Buried slide"** technique for studying soil micro flora.

Van Niel (1931USA) studied chemoautotrophic bacteria and bacterial photosynthesis.

Bortels (1936) demonstrated the importance of molybdenum in accelerating nitrogen fixation by nodulating legumes.

Garrett (1936) established the school in UK on «Soil fungi and ecological classification».

Kubo (1939, Japan) showed/proved-the role and importance of "leghaemoglobin" (Red pigment) present in root nodules of legumes in nitrogen fixation.

Ruinen (1956) Dutch microbiologist coined the term **"Phyllosphere"** to denote the region of leaf influenced by microorganisms.

Alien et al (1980) (suggested that VAM fungi stimulate plant growth by physiological effects other than by enhancement of nutrient uptake.

Jensen (1942) developed the method of studying nodulation on agar media in test tubes.

Barbara Mosse and J. W. Gerdemann (1944) reported occurrence of VAM (vesicular-arbuscular Mycorrhiza) fungi (*Glomus, Aculopora genera*) in the roots of agricultural crop plants which helps in the mobilization of phosphate.

Starkey (1945) studied role of bacteria (Bacillus and Clostridium) in the transformation of iron.

Barker (1945) studied anaerobic fermentation by methane bacteria (*Methanococcus, Methanosarcina*)

Thornton, (1947), studied root nodule bacteria form clovers.

Virtanen (1947) studied chemistry and mechanism of leghaemoglobin in nitrogen fixation.

Nutman (1948 England) studied hereditary mechanism of root nodulation in legumes.

Burris and Wilson (1957) developed the **"Isotope technique"** to quantify the amount of nitrogen fixed and further isolated and characterized the enzyme "Nitrogenase".\

Bergersen (1957 Australia) elaborated the biochemistry of nitrogen fixation in legume root nodules.

Carnham (1960 USA) discovered nitrogen fixation by cell-free extract of *Clostridium pasteurianum.*

Alexander Fleming (1881 – 1955)

In 1928, while working on influenza virus, he observed that mould had developed accidentally on a staphylococcus culture plate and that the mould had created a bacteria-free circle around itself. He was inspired to further experiment and he found that a mould culture prevented growth of staphylococci, even when diluted 800 times. In 1929 he discovered the Antibiotic *Penicillin,* which is the important milestone in medical microbiology. Florey and Chain later isolated penicillin in pure culture. Alexander Fleming got the Nobel Prize in medicine in the year 1945. He found that natural substances / natural products are having antimicrobial activity. He reported that Nasal mucous, saliva are having antimicrobial property due to the action of lysozyme. He started the **"School of soil microbiology"** at Cornell University to study microbial aspects of pesticides degradation (1961) and developed the antibiotic «Penicillin» from the fungus *Penicillium notatum* (1929).

Date, Brockwell and Roughley (1962, Australia) developed the technique of bio-inoculants production & seed application.

Hardy & Associates (1968, USA) developed the technique of measurement of nitrogenase activity by acetylene-reduction test coupled with gas chromatography and thereby estimation of biological nitrogen fixation.

R J Swaby (1970, Australia) developed «**Biosuper**" containing rock phosphate sulphur and Thiobacillus which was used to enhance the phosphorus nutrition of plants.

Foog and Stewart (1970, UK) intensified the work on N2 fixing blue-green algae.

Trinick (1973, Australia) isolated Rhizobia from root nodule of genus Trema (Parasponia) which was an unique association of Rhizobium with non-leguminous plants causing root nodulation.

Dobereiner and associates (1975, Brazil) studied nitrogen fixing potential of Azospirillum in some tropical forage grasses like Digitaria, Panicum and some cereals like maize, sorghum, wheat, rye *etc.* in their roots. He reported four species of Azospirillum *viz. A. lipoferum, A. brasilense, A. amazonense* and *A. serpedica*. He coined the term **"Associative Symbiosis"** to denote the association between nitrogen fixing Azospirillum and cereal roots. Recently this terminology has been changed and renamed as "Diazotrophic Biocoenocis".

Challham and Associates (1978) isolated an actinomycetous endophyte *Frankia sp* from root nodules of *Camptonia peregrina* which is again an example of non-leguminous root nodulation.

Dommergues & associates (France and Senegal) had discovered / reported nodules on stem of *Sesbania rostrata* which could fix nitrogen and therefore this legume can be used as an excellent green manure crop in low land rice cultivation. Similarly they also discovered N2 fixing stem nodules on *Casurina sp* caused by Frankia, an actinomycete.

Louis Pasteur Proved the role of soil microorganisms in biochemical changes of elements. He also showed that decomposition of organic residues in soil was dependent on the nature of organic matter and environmental conditions.

Brefeld Introduced the practice of isolating soil fungi by **"Single Cell"** technique and cultivating / growing them on solid media. He used gelatin (first solidifying agent) in culture media as solidifying agent.

Gerretsen & Mulder (Holland) studied **"Phosphate mobilization"** by soil microorganisms and showed the importance of molybdenum in nitrogen metabolism by microorganisms.

Fritch, fogg & Stewart (UK) and lyengar (India) studied fixation by algae in general and micro algae in particular. They also intensified the work on N2 fixing BGA

James Trappe and Don Marx worked on ectomycorrhiza, colonizing the roots of forest trees.

W. S. Cook, G. C. Papavizas, J. Baker and N.S. Kerr contributed to the field of biological control of plant pathogens using antagonistic organisms from soil. From the beginning of 20th century emphasis was given to the study of microorganisms in soil in relation to their physiology, ecology, interrelationship, role in soil processes and soil fertility. Further role of fungi and actinomycetes in cellulose decomposition was better understood and cellulose decomposing, sulphur oxidizing, iron bacteria *etc* were isolated from soil and studied in detail.

History of Soil Microbiology in India

During last few decades greater emphasis has been given on some of the important aspects in soil microbiology in India which are:

1. Characterization of N_2 fixing Azotobacter, Rhizobium, Beijerinckia, BGA *etc*.
2. Studies on P- solubilizing bacteria and fungi, celluloytic microorganisms, silage production role of humic acid *etc*.
3. Establishment (1979) of All-India Coordinated Project (AICP) on BNF at IARI and field oriented work on BNF.
4. Standardization of methods of bio-inoculants application to seed and soil.
5. Seed bacterization and response of crops to bio-inoculants.

Some of the most important contributions made on the different aspects in the field of soil microbiology by the scientists and research institutes in the country are highlighted in the following paragraphs:

C. N. Acharya (1940) contributed towards the better utilization of Agricultural wastes for the production of biogas & compost.

Sundara Rao (1962) established the **"Division of Microbiology"** at IARI New Delhi.

Madhok (Punjab) introduced the practice of using bacterial cultures for berseem.

Sanyasi Raju & Rajagopalan (Coimbatore) initiated the research work on root nodulation in legumes at Madras, Agil. College.

P. K. Dey (West Bengal) worked on free living N2 fixing organisms *viz Azotobacter, Beijerinckia* and BGA in rice fields and discovered N2 fixation by BGA in paddy.

M.O.P. lyengar (Madras Univ.) laid foundation stone of algal research in India.

Sadasivan (Madras) and Saxena (Allahabad) studied ecology and physiology of soil fungi along with rhizosphere phenomenon.

Singh B.N. pioneering research on soil protozoa in India.

Bhar J.V. (Bangalore) initiated work on the role of earthworms in the maintenance of soil fertility, biological nitrogen fixation and microbiology of phyllosphere.

Thirumalacher (Hindustan Antibiotics, Pune) developed antifungal antibiotics like Haymycin and Aureofungin.

Nandi (Bose Res. Institute, Calcutta) worked on production technology of antibiotics and bacterial fertilizers (Biofertilizers).

Desikachray (Madras) studied taxonomy of BGA in India.

Thomas (BARC, Mumbai) studied physiology of algae in India.

Raja Rammohan Rao (CRRI, Cuttak) studied on rhizosphere nitrogen fixation phenomenon.

Bhagyaraj (GKVK, Bangalore) studied Mycorrhiza and N2 fixation interactions.

Verma (JNU, Delhi) studied / worked on sulphur metabolism.

Subramaniam & Mahadevan (Univ. Madras) studied fundamental aspects of N2 fixation.

Modi, Sushil Kumar, Das and Thomas carried research on "Genetics of "Nif" gene in relation to BNF by *Rhizobium, Azospirillum* and *Kelbsiella.*

Bharadwaj (Palampur) studied / worked on microbiology of organic matter decomposition & role of celluloytic microorganisms.

Gaur (IARI) and Mishra (Hissar) studied the role of celluloytic microorganisms in accelerating the process of composting and compost making.

Karla and Garcha (Ludhiana) studied the phenomenon of cellulose degradation and legume bacteriology.

Ranganathan & Nellakantan (NDRI, Karnal) worked on silage microbiology and process of anaerobic decomposition in biogas production.

Vadher, Gupta, Sethunatathan and Raghu studied role of soil enzymes and microbiology of pesticide degradation in soil.

Dart & Wani (non symbiotic N_2 fixation), **Thomas, Kumar Rao, Nambiar and Rupela** (symbiotic N_2 fixation) and **Krishna** (VAM fungi), these scientist at ICRISAT, Hyderabad work on symbiotic & non-symbiotic N_2 fixation in Gram, Groundnut, Arhar, Sorghum and Millets.

N. V. Joshi (1920) reported first isolation and identification *of Rhizobium* from different cultivated legumes

Gangulee and Madhok, studied physiology of *Rhizobium* and production of *Rhizobium* inoculants.

Sen and Pal (1957) studied solubilization of phosphate by soil microorganisms.

A. Sankaran (1958) standardized quality of legume inoculants for first time in India.

P. K. Dey and R. Bhattacharya isolated for the first time a new, non-symbiotic N2 fixing bacterium *Derixa gummosa* in the world.

V. Iswaran (1959) reported the use of Indian peat as carrier for Biofertilizers production.

Dube J. N. (1975) reported coal (wood-coal), an alternative to peat as carrier material for biofertilizer production.

3

Distribution and Importance of Microorganisms

Soil is a bioorganic - mineral complex of the loose surface layer of the earth's crust. Soil microbiology is the science that deals with the millions of microscopic forms of life found in the soil. It is stated that one hectare - foot (one hectare to a depth of 30cm) contains 2 to 10 metric tons of living microorganisms. Soil microorganisms are in a constant struggle for existence and are believed to be in a state of dynamic equilibrium. Without microorganisms, soil would be an inactive geological mass incapable of supporting plant life and without plant life, animal life would not be possible. This soil population consists mainly of bacteria, fungi, actinomycetes, algae and protozoa. The soil ecosystem includes these microbial groups as well as the inorganic and organic constituents of a given site. The collections of cells represented in the community are considered as distinct populations. All the inhabitants of the particular locality make up the community.

Types of Microorganisms in Soil

Living organisms both plants and animals, constitute an important component of soil. The pioneering investigations of a number of early microbiologists showed for the first time that the soil was not an inert static material but a medium pulsating with life. The soil is now believed to be a dynamic or rather a living system, containing a dynamic population of organisms/microorganisms. Cultivated soil has relatively more population of microorganisms than the fallow land, and the soils rich in organic matter contain much more population than sandy and eroded soils. Microbes in the soil are important to us in maintaining soil fertility / productivity, cycling of nutrient elements in the biosphere and sources of industrial products such as enzymes, antibiotics, vitamins, hormones, organic acids *etc*. At the same time certain soil microbes are the causal agents of human and plant diseases. The soil organisms are broadly classified in to two groups *viz* soil flora and soil fauna, the detailed classification of which is as follows.

Soil Organisms

A. Soil Flora

a) Microflora: 1. Bacteria 2. Fungi, Molds, Yeast, Mushroom 3. Actinomycetes 4.Algae.

b) **Macroflora:** Roots of higher plants

B. Soil Fauna

a) **Microfauna:** Protozoa, Nematodes

b) **Macrofauna**: Earthworms. moles, ants & others.

As soil inhabit several diverse groups of microorganisms, but the most important amongst them are: bacteria, actinomycetes, fungi, algae and protozoa. The characteristics and their functions / role in the soil are described in the next topics.

I. Bacteria

Bacteria are the most dominant group of M.O in soil and more numerous than the other four combined. They present in all types of soil but their population decreases as the depth of soil increases (Horizon A > B > C). The number of cells of bacteria in the soil is always great, but the individuals are small, (µm in length). Because of the minute size of the bacteria and large cells or filaments of the other 4 groups, the bacteria probably account for appreciably less than half of the total microbial cell mass. In aerated soil, B, will be dominating, alone are responsible for all the activities in environment lacking O_2 or little O_2. In transformation process Bacteria stand first, due to their rapid growth and capacity of vigorous decomposition of variety of substrates.

Soil microbiological population has been divided into two broad groups. 1. Autohcthonous 2. Zymogenous.

Autochthonous or native microbes; Indigenous, which are characteristic of the particular soil and always found to be there.*e.g.* Arthrobacter

Zymogenous, or fermentative organisms require an external source of and their normal population in soil is low (Pseudomonas, Bacillus). When specific substrates are added to soil, the population is increased. Then gradually declines when the added substrate is exhausted. *e.g.* Cellulose decomposers. N utilizing bacteria, nitrifiers *etc.*,

All the bacteria essentially need carbon, energy and electron for their normal cellular activity, growth and reproduction. Based on the source of above, bacteria can be classified in to following groups.

1. **Autotrophs:** They utilize CO_2 as sole source for carbon requirement. Autotrophic bacteria are capable synthesizing their food from simple inorganic nutrients, while heterotrophic bacteria depend on pre-formed food for nutrition. All autotrophic bacteria utilize CO_2 (from atmosphere) as carbon source and derive energy either from sunlight (photoautotrophs, *e.g.* Chromatrum. Chlorobium. Rhadopseudomonas or from the oxidation of simple inorganic substances present in soil (chemoautotrophs *e.g.* Nitrobacter, Nitrosomonas, Thiaobacillus).
2. **Heterotrophs:** They derive their carbon from organic compounds.Majority of soil bacteria are heterotrophic in nature and derive their carbon and energy from complex organic substances/organic matter, decaying roots and plant residues. They obtain their nitrogen from nitrates and ammonia

compounds (proteins) present in soil and other nutrients from soil or from the decomposing organic matter. Certain bacteria also require amino acids, B- Vitamins, and other growth promoting substances also.*e.g*. Symbiotic & Non - Symbiotic N_2 fixers, Ammonifier, Cellulose Decomposers, Denitrifiers

3. **Phototrophs:** Bacteria capable of deriving energy from sunlight.
4. **Chemotrophs:** They utilize inorganic chemicals for their energy.
5. **Lithotroph:** The organisms utilize inorganic substances as electron source.
6. **Organotroph:** Bacteria capable of utilizing organic substances as electron source.

They may exist in combinations of the above groups as follows. Photolithotrophs; Chemolithotrophs; Photoorganotrops; Chemoorganotrophs; Photoautotrophs; Photoheterotrophs; Chemoautotrophs; Chemoheterotrophs.

Abundance of Bacteria

In Arctic regions where in the frozen condition 9-10 months / year bacterial counts of 10^6 / g are observed even when temperature remains below freezing point. Such organisms are in a state of dormancy awaiting the spring to recover. Desert soil is another extreme. Bacteria exist even in oven dry state. Spore forming bacilli are often predominant in this environment. Common soil bacteria are Arthrobacter, Bacillus, Clostridium, Micrococcus.

As per the system proposed in the Bergey's Manual of Systematic Bacteriology, most of the bacteria which are predominantly encountered in soil are taxonomically included in the three orders, Pseudomonadales, Eubacteriales and Actinomycetales of the class Schizomycetes. The most common soil bacteria belong to the genera Pseudomonas, Arthrobacter, Clostridium Achromobacter, Sarcina, Enterobacter *etc*. The another group of bacteria common in soils is the Myxobacteria belonging to the genera Micrococcus, Chondrococcus, Archangium, Polyangium, Cyptophaga.

Importance

1. In transformation processes, bacteria stand first due to their rapid growth and capacity of vigorous decomposition of variety of substrates.
2. Most of the organisms are involved in N_2 fixation, P solubilization, CO_2 fixation, S and Fe transformation, Si solubilization *etc*.,
3. They also act as molecular vectors.

Functions / Role of Bacteria

Bacteria bring about a number of changes and biochemical transformations in the soil and thereby directly or indirectly help in the nutrition of higher plants growing in the soil. The important transformations and processes in which soil bacteria play vital role are: decomposition of cellulose and other carbohydrates, ammonification (proteins ammonia), nitrification (ammonia-nitrites-nitrates), denitrification (release of free elemental nitrogen), biological fixation of atmospheric

4. They also serve as food for bacteria.
5. Certain fungi belonging to sub-division Zygomycotina and Deuteromycotina are predaceous in nature and attack on protozoa & nematodes in soil and thus, maintain biological equilibrium in soil.
6. They also plays important role in soil aggregation and in the formation of humus.
7. Some soil fungi are parasitic and causes number of plant diseases such as wilts, root rots, damping-off and seedling blights *e.g. Pythium, Phyiophlhora, Fusarium, Verticillium* etc.
8. Number of soil fungi forms mycorrhizal association with the roots of higher plants (symbiotic association of a fungus with the roots of a higher plant) and helps in mobilization of soil phosphorus and nitrogen *e.g.* Endomycorrhiza - Glomus, *Gigaspora, Aculospora,* and Ectomycorrhiza - *Amanita, Boletus, Entoloma, Lactarius.*

III. Actinomycetes

They are unicellular like bacteria, but produce a mycelium which is non-septate (coenocytic) and more slender, tike true bacteria they do not have distinct cell-wall and their cell wall is without chitin and cellulose (commonly found in the cell wall of fungi). On culture media unlike slimy distinct colonies of true bacteria which grow quickly, actinomycetes colonies grow slowly, show powdery consistency and stick firmly to agar surface. They produce hyphae and conidia / sporangia like fungi. Certain actinomycetes whose hyphae undergo segmentation resemble bacteria, both morphologically and physiologically.

A transitional group between the simple Bacteria and Fungi are the actinomycetes. These are microorganisms producing slender, branded filaments that develop into a mycelium in all soil. They are classified under bacteria, but resemble fungi in some aspects.

- Produces mycelium with extensive branching
- Like Fungi many actinomycetes form aerial mycelium and conidia
- Growth in liquid medium does not resemble (which bacteria is turbid), but form clumps and pellets like Fungi.

Certain actinomycetes resemble Mycobacterium and Corynebacterium in all respects both morphologically and physiologically including susceptibility to virus attack.

Actinomycetes differ from fungi that they do not have chitin and cellulose which are commonly found in the cell walls of fungi.The colony characters are not similar to bacteria.Some species have flagella that resemble those of true Bacteria.Similarities in cell wall composition and sensitivity to antibacterial compounds. Actinomycetes are sensitive to antibacterial compounds and not antifungal compounds

Common genera in soils are Streptomyces (70%), Nocardia, Micromonaspora, Frankia. They elaborate musty odour, an odour reminiscent of freshly turned soil.

The metabolite of Streptomyces formed geosmin and other volatile products, which is responsible for the characteristic earth odor/ smell.

Actinomycetes belonging to the order of Actinomycetales are grouped under four families *viz* Mycobacteriaceae, Actinomycetaceae, Streptomycetaceae and Actinoplanaceae. Actinomycetous genera which are agriculturally and industrially important are present in only two families of Actinomycetaceae and Strepotmycetaceae.

Distribution and abundance

Actinomycetes are numerous and widely distributed not only in soil but also in a variety of other habitats including composts, river mud and lake bottoms. Present in surface soil,compost and also in lower horizons to considerable depth. In abundance they are second only to the Bacteria and the viable counts almost equal to both. Saprophytic existence, but a few species can cause diseases of plants, domestic animals and even humans. Plate count estimates give values ranging from 10^4 to 10^8 per gram of soil. They are sensitive to acidity / low PH (optimum PH range 6.5 to 8.0) and waterlogged soil conditions.

The population of actinomycetes increases with depth of soil even up to horizon 'C' of a soil profiles. They are heterotrophic, aerobic and mesophilic (25-30°C) organisms and some species are commonly present in compost and manures are thermophilic growing at 55-65° c temperature (*e.g.* Thermoatinomycetes, Streptomyces).Their population is 10^5 to 10^8 / g in temperate zone, but lower counts in waterlogged soils, acid peat, arctic, Tundra regions. In alkaline areas, especially when dry, the relative abundance is high.

In the order of abundance in soils, the common genera of actinomycetes are Streptomyces (nearly 70%), Nocardia and Micromonospora although Actinomycetes, Actinoplanes, Micromonospora and Streptosporangium are also generally encountered.

Importance

- ❑ Involved in the decomposition of resistant components of plant and animal tissues.
- ❑ They play active role in the transformations at high temperature particularly in the manure and compost pits (as thermophiles are dominant).
- ❑ Cause certain soil borne diseases of plants, potato scab of apple, sweet potato pox.
- ❑ Some actinomycetes cause infection to humans, animals.
- ❑ Importance in microbial antagonism by production of antibiotics and production of enzymes.
- ❑ Utilized in the antibiotic production at a larger level.

Functions / Role of Actinomycetes

1. Degrade/decompose all sorts of organic substances like cellulose, polysaccharides, protein fats, organic-acids *etc.*

2. Organic residues / substances added soil are first attacked by bacteria and fungi and later by actinomycetes, because they are slow in activity and growth than bacteria and fungi.
3. They decompose / degrade the more resistant and indecomposable organic substance/matter and produce a number of dark black to brown pigments which contribute to the dark colour of soil humus.
4. They are also responsible for subsequent further decomposition of humus (resistant material) in soil.
5. They are responsible for earthy / musty odor / smell of freshly ploughed soils.
6. Many genera species and strains (*e.g.* Streptomyces if actinomycetes produce/synthesize number of antibiotics like Streptomycin, Terramycin, Aureomycin *etc.*
7. One of the species of actinomycetes *Streptomyces scabies* causes disease "Potato scab" in potato.

IV. Yeast

Yeasts are non-filamentous fungi. Their presence may be demonstrated in most of the soils. Common soil yeasts are Candida, Debaryomyces, Rhodotorula, Torulopsis. Yeasts have been found in comparable numbers in soils of Antarctica, grasslands, cultivated fields, and forests.

Function and Activity

- Involved in the degradation of complex molecules.
- They can utilize and degrade the major plant constituents – cellulose, hemicellulose, pectin, starch and lignin.
- Participate in humus formation from fresh organic residues.
- Carry out inorganic transformations and influences the formation of stable aggregates by means of hyphal penetration and the mechanical binding of the particles.

V. Algae

Algae are present in most of the soils where moisture and sunlight areavailable. Their number in soil usually ranges from 100 to 10,000 per gram of soil. They are photoautotrophic, aerobic organisms and obtain CO_2 from atmosphere and energy from sunlight and synthesize their own food. They are unicellular, filamentous or colonial. Soil algae are divided in to four main classes or phyla as follows:

1. Cyanophyta (Blue-green algae)
2. Chlorophyta (Grass-green algae)
3. Xanthophyta (Yellow-green algae)
4. Bacillariophyta (diatoms or golden-brown algae)

Out of these four classes / phyla, blue-green algae and grass-green algae are more abundant in soil. The green-grass algae and diatoms are dominant in the soils of temperate region while blue-green algae predominate in tropical soils. Green-algae prefer acid soils while blue green algae are commonly found in neutral and alkaline soils. The most common genera of green algae found in soil are: Chlorella, Chlamydomonas, Chlorococcum, Protosiphon *etc.* and that of diatoms are Navicula, Pinnularia. Synedra, Frangilaria.

Blue green algae are unicellular, photoautotrophic prokaryotes containing Phycocyanin pigment in addition to chlorophyll. They do not posses flagella and do not reproduce sexually. They are common in neutral to alkaline soils. The dominant genera of BGA in soil are: Chrococcus, Phormidium, Anabaena, Aphanocapra, Oscillatoria *etc.* Some BGA posses specialized cells know as **"Heterocyst"** which is the sites of nitrogen fixation. BGA fixes nitrogen (non-symbiotically) in puddle paddy/water logged paddy fields (20-30 kg/ha/season). There are certain BGA which possess the character of symbiotic nitrogen fixation in association with other organisms like fungi, mosses, liverworts and aquatic ferns Azolla, eg Anabaena-Azolla association fix nitrogen symbiotically in rice fields.

They are photosynthetic organisms and abundant in habitats, where adequate moisture and accessible light are available. Isolation can be done in liquid media and by plating with agar media. Because of their similarities to bacteria, the blue green algae (cyanobacteria) are classified together with bacteria.

Photoautotrophic nutrition

In the absence of light heterotrophy occur in some species of Chlorophyta, Cyanrphyta and Diatoms. These species use the oxidation of organic carbon to replace the light in supplying energy for anabolic process. (Facultative photoautotrophs)

Occurrence of this photosynthetic microorganism has been recorded in soils throughout the world. Algae are moderately adaptable to environmental changes; persist in unfavourable circumstances such as in alkaline and desert soils. Flooded paddy field provides conducive environment for the growth of algae. The common inhabitants are Anabaena, Nostoc, Calothrix and Oscillatoria.

Importance

- Involved in CO_2 fixation
- Colonizes the barren surface and corrode and weather rocks which contribute to soil formation
- Algal layer covers the rocks, on death it favors secondary colonizers
- Surface algal bloom reduces erosion losses probably by binding together with soil particles
- Improve soil structure, texture and add fertility to soil after decay
- Photosynthesis liberates 02 , provides 02 to the roots and other microbes
- N2 fixation and ammonia excretion
- N2 is released after decomposing of algal biomass

- ❑ Production of metabolites
- ❑ Single cell protein and β carotene production

Functions / Role of Algae or BGA

1. Plays important role in the maintenance of soil fertility especially in tropical soils.
2. Add organic matter to soil when die and thus increase the amount of organic carbon in soil.
3. Most of soil algae (especially BGA) act as cementing agent in binding soil particles and thereby reduce/prevent soil erosion.
4. Mucilage secreted by the BGA is hygroscopic in nature and thus helps in increasing water retention capacity of soil for longer time/period.
5. Soil algae through the process of photosynthesis liberate large quantity of oxygen in the soil environment and thus facilitate the aeration in submerged soils or oxygenate the soil environment.
6. They help in checking the loss of nitrates through leaching and drainage especially in un-cropped soils.
7. They help in weathering of rocks and building up of soil structure.

VI. Protozoa

The subterranean earth contains protozoa, earthworms, nematodes, and insects. Most abundant one is protozoa, the simple form of animal life. Most of the soil forms are flagellates, amoebae or ciliates. Primitive, unicellular and majority of them are dependent on preformed organic matter for their nutrition from soluble organic and inorganic substances, or by phagotrophic nutrition (direct feeding upon microbial cells). They are regulating the biological equilibrium in soil by engulfing the bacteria. They also participate in the decomposition of plant remains.

These are unicellular, eukaryotic, colourless, and animal like organisms (Animal kingdom). They are larger than bacteria and size varying from few microns to a few centimeters. Their population in arable soil ranges from 10,000 to 1,00,000 per gram of soil and are abundant in surface soil. They can withstand adverse soil conditions as they are characterized by **"cyst stage"** in their life cycle. Except few genera which reproduce sexually by fusion of cells, rest of them reproduces asexually by fission / binary fission. Most of the soil protozoa are motile by flagella or cilia or pseudopodia as locomotors organs.

Depending upon the type of appendages provided for locomotion, protozoa are

1. Rhizopoda (Sarcondia)
2. Mastigophora
3. Ciliophora (Ciliata)
4. Sporophora (not common Inhabitants of soil)

Class-Rhizopoda: Consists of protozoa without appendages usually have naked protoplasm without cell-wall, pseudopodia as temporary locomotory organs are present some times. Important genera are Amoeba, Biomyxa, Euglypha, *etc.*

Class Mastigophora: Belongs to flagellated protozoa, which are predominant in soil. Important genera are: Allention, Bodo, Cercobodo, Cercomonas, Entosiphon Spiromonas, Spongomions and Testramitus. Many members are saprophytic and some posses chlorophyll and are autotrophic in nature. In this respect, they resemble unicellular algae and hence are known as "Phytoflagellates".

The soil protozoa belonging to the class ciliate / ciliophora are characterized by the presence of cilia (short hair-like appendages) around their body, which helps in locomotion. The important soil inhabitants of this class are Colpidium, Colpoda, Balantiophorus, Gastrostyla, Halteria, Uroleptus, Vortiicella, Pleurotricha *etc.*

Protozoa are abundant in the upper layer (15 cm) of soil. Organic manures protozoa. Soil moisture, aeration, temperature and PH are the important factors affecting soil protozoa.

Function / Role of Protozoa

1. Most of protozoans derive their nutrition by feeding or ingesting soil bacteria belonging to the genera *Enterobacter, Agrobacterium, Bacillus, Escherichia, Micrococcus,* and *Pseudomonas* and thus, they play important role in maintaining microbial / bacterial equilibrium in the soil.
2. Some protozoa have been recently used as biological control agents against phytopathogens.
3. Species of the bacterial genera *viz. Enterobacter* and *Aerobacter* are commonly used as the food base for isolation and enumeration of soil protozoans.
4. Several soil protozoa cause diseases in human beings which are carried through water and other vectors, *e.g.* Amoebic dysentery caused by *Entomobea histolytica*

Importance of soil microorganisms

Soil microorganisms are very important as almost every chemical transformation taking place in soil involves active contributions from these organisms. In particular, they play an active role in soil fertility as a result of their involvement in the cycle of nutrients like carbon and nitrogen, which are required for plant growth. For example, soil microorganisms are responsible for the decomposition of the organic matter entering the soil and therefore in the recycling of nutrients. Certain soil microorganisms such as mycorrhizal fungi can also increase the availability of mineral nutrients (*e.g.* phosphorus) to plants. Other soil microorganisms can increase the amount of nutrients present in the soil. For instance, nitrogen-fixing bacteria can transform nitrogen gas present in the soil atmosphere into soluble nitrogenous compounds that plant roots can utilize for growth. These microorganisms, which improve the fertility status of the soil and contribute to plant growth, have been termed 'biofertilizers' and are receiving increased attention for use as microbial inoculants in agriculture. Similarly, other soil microorganisms have been found to produce compounds (such as vitamins and plant hormones) that can improve plant health and contribute to higher crop yield. These microorganisms (called 'phytostimulators') are also studied for possible use as microbial inoculants to improve crop yield.

In contrast to these beneficial soil microorganisms, other soil microorganisms are pathogenic to plants and may cause considerable damage to crops. Large numbers of pathogenic microorganisms are routinely found in the soil and many of them can infect the plant through the roots. However, certain native microorganisms present in the soil are antagonistic to these pathogens and can prevent the infection of crop plants. Antagonism against plant pathogens usually involves competition for nutrients and/or production of inhibitory compounds such as secondary metabolites (antimicrobial metabolites and antibiotics) and extracellular enzymes. Other soil microorganisms produce compounds that stimulate the natural defense mechanisms of the plant and improve its resistance to pathogens. Collectively, these soil microorganisms have been termed as 'biopesticides' and represent an emerging and important alternative (*i.e.* biological control) to the use of chemical pesticides for the protection of crops against certain pathogens and pests.

Factors affecting microbial activities in soil

Several environmental conditions affect the density and composition of the microflora and frequently alter their activities in soil. Primary factors include soil moisture, aeration, temperature, pH, organic matter and inorganic fertilizers. Lesser variables are the secondary factors, which include crop rotation, season, soil depth, cultivation practices *etc.*

I. Primary factors

Soil moisture

Soil moisture is one of the important factors influencing the microbial population. Water is the major component of protoplasm; an adequate supply must be available for vegetative development of organisms. But, when it becomes excessive, proliferation is suppressed due to limitation in gaseous exchange and lowers the availability of O_2 supply creating anaerobic environment. Most of the organisms prefer a moisture percentage between 20 and 60 per cent. Many bacteria and fungi are able to adjust themselves to different moisture conditions. Under dry conditions, the bacteria may form spores that can resist the drought conditions. The fungi may sporulate or form chlamydospores to tide over adverse conditions. The protozoans also may form cysts and can survive under dry conditions. Actinomycetes are the chief group of organisms that prefer dry conditions. It is believed that the concentration of nutrients are diluted and also the aeration is very much limited at moisture and hence only the anaerobic and microaerophilic organisms can develop better.

Soil air

It is directly linked up the moisture level of the soil. Most of the forms are active in aerated soil and under waterlogged soils anaerobic and microaerophilic forms develop.

Soil temperature

It exhibits considerable influence on the microbial population. Though microorganisms have been found to exist under extreme temperature conditions, such as – 60° C and + 60° C the soil temperature usually does not reach such extremes. Microbial population varies both quantitatively and qualitatively under extreme conditions. In tropical and subtropical regions, temperatures vary widely in summer and winter seasons and the population may also be varied. In temperate regions, there is not much variation in temperatures of seasons and not much variation in the soil population. Soil temperature influences the temperature of air, water and solid phase of the soil. Soil water and temperature exert a combined influence on the microbial population.

Soil organic matter

Community size is related to the organic matter content, so that humus rich localities have the largest biological numbers. Organic matter content varied with soil types from less than 0.5 per cent in desert soils to 40 per cent in peaty soils. Organic matter being the chief source of energy and food for most soil organisms, it has great influence on the population. Nature of Organic matter is responsible for the differential stimulation of the population. There are several indirect effects of the organic matter on soil microflora. It influences the structure and texture of soil, besides enriching with nutrients for plants and microorganisms. Such influences on the soil also greatly affect the activity of the soil microorganisms.

Soil pH

It is a key factor influencing the microflora of soil. It influences enzyme system and thus plays an important role in microbial activity. In general, fungi thrice better than bacteria and actinomycetes in acid soils. Bacteria flourish well in neutral and alkaline soils. The saline and alkaline soils have different microflora. Salinity is due to excess salts; alkalinity is due to high H-ion concentration. Several direct and indirect effects of H-ion and salt concentration in soil are exerted on microbial population.

Fertilizer application

Application of fertilizers, to the soil improves the microbial activity because of the availability of more readily obtainable nutrients. Some fertilizers may however have inhibitory effect on specific bacterial types. Continuous application of ammoniacal fertilizers favours the growth of fungi due to the formation of nitric acid that inhibits the growth of bacteria and actinomycetes. Addition of nitrate inhibits the activity of free living N_2 fixing bacteria like Azotobacter. Some of the autotrophs are encouraged by the addition of fertilizers.

Cropping and vegetation

The crop plants exhibit two kinds of effects. One is through root exudates, which may have different compounds with reference to the crops grown. Vegetation has selective stimulation over population. Continuous cultivation leads to more microbial activity than the uncultivated land.

II. Secondary factors

Crop rotation

Crop rotation with different species like legumes, graminaceous plants, *etc.* brings about different stimulatory effect on the microflora. Some crops have deeper roots than others and some are more fibrous. Such variations bring about physical changes in soil, which in turn may have direct and indirect stimulatory effects on the soil microflora.

Cultural practices

Various cultural practices, such as tillage operations and irrigation have several physical and chemical changes in soil, which are reflected on the soil microflora. Through subsoil ploughing deeper layers may get better aeration and there may be quicker multiplication of aerobic organisms. Weeding, irrigation *etc.* may influence the microbial populations in the soil.

Soil depth

Most of the organisms are almost in top layers, largely in upper few centimeters and decline with greater depth, more active at few cm down and less active at deeper layers. Low O_2 and less sunlight in deeper layers reduce soil microbial population.

Season

Cell numbers are greatest during the spring and autumn and a decline occur during hot, dry and winter, the cells remain in a state of dormancy for biochemical inactivity. The influence is due to mainly the alterations in temperature, moisture as well as availability of organic matter.

III. Specific influences

- Sunlight favours algae and other autotrophs.
- Herbicide application has devasting effect on algae.
- Attack by neighbours – protozoa, nematode, earthworms consume algae.
- Presence of parasites and predators – *e.g.* viruses eat on specific bacteria
- Increase in bacterial population increases protozoans also.

IV. Other factors

Burning of topsoil

- Leads to partial sterilization of the top soil and may kill protozoan population, which may lead to the increase in bacterial population. This condition may affect biological equilibrium in soil.
- Application of nematicide, fungicide and bactericide may exhibit partial sterilization.Thus, every change in crop production directly or indirectly alters the soil microflora, still many more are not clearly understood.

Factors Affecting Distribution and Population of Soil Microorganisms

Soil microorganisms (Flora & Fauna), just like higher plants depends entirely on soil for their nutrition, growth and activity. The major soil factors which influence the microbial population, distribution and their activity in the soil are

1. Cultural practices
2. Soil fertility
3. Soil moisture
4. Soil temperature
5. Soil aeration
6. Light
7. Soil pH (H-ion Concentration)
8. Organic matter
9. Food and energy supply
10. Nature of soil
11. Microbial associations.

All these factors play a great role in determining not only the number and type of organism but also their activities. Variations in any one or more of these factors may lead to the changes in the activity of the organisms which ultimately affect the soil fertility level. Brief account of all these factors influencing soil micro flora / organisms and their activities is activities are discussed paragraphs.

1. Cultural practices (Tillage): Cultural practices *viz.* cultivation, crop rotation, application of manures and fertilizers, liming and gypsum application, pesticide/ fungicide and weedicide application have their effect on soil organism. Ploughing and tillage operations facilitate aeration in soil and exposure of soil to sunshine and thereby increase the biological activity of organisms, particularly of bacteria. Crop rotation with legume maintains the favorable microbial population balance, particularly of N2 fixing bacteria and thereby improve soil fertility.

- Liming of acid soils increases activity of bacteria and actinomycetes and lowers the fungal population. Fertilizers and manures applied to the soil for increased crop production, supply food and nutrition not only to the crops but also to microorganisms in soil and thereby proliferate the activity of microbes.
- Foliar or soil application of different chemicals (pesticides, fungicides, nematicides *etc.*) in agriculture are either degraded by the soil organisms or are liable to leave toxic residues in soil which are hazardous to cause profound reduction in the normal microbial activity in the soil.

2. Soil fertility: Fertility level of the soil has a great influence on the microbial population and their activity in soil. The availability of N, P and K required for plants as well as microbes in soil determines the fertility level of soil. On the other hand soil micro flora has greater influence on the soil fertility level.

3. Soil moisture: It is one of the important factors influencing the microbial population & their activity in soil. Water (soil moisture) is useful to the microorganisms in two ways *i.e.* it serve as source of nutrients and supplies hydrogen / oxygen to the organisms and it serve as solvent and carrier of other food nutrients to the microorganisms. Microbial activity & population proliferate best in the moisture range of 20% to 60%. Under excess moisture conditions / water logged conditions due to lack of soil aeration (Oxygen) anaerobic microflora become active and the aerobes get suppressed. While in the absence of adequate moisture in soil, some of microbes die out due to tissue dehydration and some of them change their forms into resting stages spores or cysts and tide over adverse conditions. Therefore optimum soil moisture (range 20 to 60 %) must be there for better population and activity of microbes in soil.

4. Soil temperature: Next to moisture, temperature is the most important environmental factor influencing the biological physical & chemical processes and of microbes, microbial activity and population in soil. Though microorganisms can tolerate extreme temperature (such as - 60 ° or + 60 u) conditions, but the optimum temperature range at which soil microorganisms can grow and function actively is rather narrow.

- Depending upon the temperature range at which microorganisms can grow and function, are divided into three groups *i.e.* psychrophiles (growing at low temperature below 10 °C) Mesophiles (growing well in the temp range of 20 ° C to 45° C) and thermopiles (can tolerate temperature above 45° C and optimum 45-60°C).
- Most of the soil microorganisms are mesophilic (25 to 40 °) and optimum temperature for most mesophiles is 37° C. True psychrophiles are almost absent in soil, and thermopiles though present in soil behaves like mesophiles. True thermopiles are more abundant in decaying manure and compost heaps where high temperature prevails.
- Seasonal changes in soil temperature affect microbial population and their activity especially in temperate regions. In winter, when temperature is low (below 50° C), the number and activity of microorganisms falls down, and as the soils warms up in spring, they increases in number as well as activity. In general, population and activities of soil microorganisms are the highest in spring and lowest in winter season.

5. Soil air (Aeration): For the growth of microorganisms better aeration (oxygen and sometimes CO_2) in the soil is essential. Microbes consume oxygen from soil air and gives out carbon dioxide. Activities of soil microbes is often measured in terms of the amount of oxygen absorbed or amount of CO_2 evolved by the organisms in the soil environment. Under high soil moisture level / water logged conditions, gaseous exchange is hindered and the accumulation of CO_4 occurs in soil air which is toxic to microbes. Depending upon oxygen requirements, soil microorganisms are grouped into categories *viz* aerobic (require oxygen for like processes), anaerobic (do not require oxygen) and microaerophilic (requiring low concentration / level of oxygen).

6. Light: Direct sunlight is highly injurious to most of the microorganisms except algae. Therefore upper portion of the surface soil a centimeter or less is usually sterile or devoid of microorganisms. Effect of sunlight is due to heating and increase in temperature (More than 45°)

7. Soil Reaction / Soil pH: Soil reaction has a definite influence / effect on quantitative and qualitative composite on of soil microbes. Most of the soil bacteria, blue-green algae, diatoms and protozoa prefer a neutral or slightly alkaline reaction between PH 4.5 and 8.0 and fungi grow in acidic reaction between PH 4.5 and 6.5 while actinomycetes prefer slightly alkaline soil reactions. Soil reactions also influence the type of the bacteria present in soil. For example nitrifying bacteria (*Nitrosomonas & Nitrobacter*) and diazotrophs like *Azotobacter* are absent totally or inactive in acid soils, while diazotrophs like *Beijerinckia, Derxia,* and sulphur oxidizing bacteria like *Thiobacillus thiooxidans* are active in acidic soils.

8. Soil Organic Matter: The organic matter in soil being the chief source of energy and food for most of the soil organisms, it has great influence on the microbial population. Organic matter influence directly or indirectly on the population and activity of soil microorganisms. It influences the structure and texture of soil and thereby activity of the microorganisms.

9. Food and energy supply: Almost all microorganisms obtain their food and energy from the plant residues or organic matter / substances added to the soil. Energy is required for the metabolic activities of microorganisms. The heterotrophs utilize the energy liberated during the oxidation of complex organic compounds in soil, while autotrophs meet their energy requirement form oxidation of simple inorganic compounds (chemoautotroph) or from solar radiation (Photoautotroph). Thus, the source of food and energy rich material is essential for the microbial activity in soil. The organic matter, therefore serves both as a source of food nutrients as well as energy required by the soil organisms.

10. Nature of Soil: The physical, chemical and physico-chemical nature of soil and its nutrient status influence the microbial population both quantitatively and qualitatively. The chemical nature of soil has considerable effect on microbial population in soil. The soils in good physical condition have better aeration and moisture content which is essential for optimum microbial activity. Similarly nutrients (macro and micro) and organic constituents of humus are responsible for absence or presence of certain type of microorganisms and their activity. For example activity and presence of nitrogen fixing bacteria is greatly influenced by the availability of molybdenum and absence of available phosphate restricts the growth of *Azotobacter.*

11. Microbial associations / interactions: Microorganisms interact with each other giving rise to antagonistic or symbiotic interactions. The association existing between one organism and another whether of symbiotic or antagonistic influences the population and activity of soil microbes to a great extent. The predatory habit of protozoa and some mycobacteria which feed on bacteria may suppress or eliminate certain bacteria. On the other hand, the activities of some of the microorganisms are beneficial to each other. For instance organic acids liberated by fungi, increase in oxygen by the activity of algae, change in soil reaction *etc.* favors the activity or bacteria and other organisms in soil.

12. Root Exudates: In the soil where plants are growing the root exudates also affects the distribution, density and activity of soil microorganism. Root exudates and sloughed off material of root surfaces provide an abundant source of energy and nutrients and thus directly or indirectly influence the quality as well as quantity of microorganisms in the rhizosphere region. Root exudates contain sugars, organic acids, amino acids, sterols, vitamins and other growth factors which have the profound effect on soil microbes.

Methods for elucidating microbial biomass of soil

More recently newer methods have been developed that use measures of components of all cells to estimate biomass. These methods include:

a) Dilution plate counts and direct microscopical counting

Dilution plates usually only are able to culture between 1 and 10% of the viable organisms in soil samples. Direct microscopic observation methods (FITC, acridine orange staining, *etc.*) usually overestimate the number of cells because they include dead organism or other particles in their count.

b) Determination of cell wall components of bacteria

Bacteria contain specific cell wall components such as muramic acid that can be released by acid hydrolysis and analyzed by High Pressure Liquid Chromatography (HPLC). The amount of muramic acid in bacterial cell walls varies depending on whether the cells are gram positive or gram negative.

c) Soil fumigation method

The fumigation of soil with $CHCl_3$ followed by extraction with various salt solutions has been used to measure microbial biomass.

d) Determination of ATP content of soils

Adenosine tri phosphate is an important energy compound in the metabolism of all living organisms. ATP is considered to be a useful indicator of life in soil. ATP can be measured with great sensitivity. Though dependent on physiological state, ATP concentrations are fairly uniform relative to cell carbon for many bacteria, algae, and protozoa. ATP is an accepted indicator of microbial biomass in aquatic and terrestrial environments. Fireflies or lightening bugs make light within their bodies. This process is bioluminescence and is shared by many other organisms, mostly sea living organisms. The cells contain a chemical called luciferin and make an enzyme called luciferase. To make light, the luciferin combines with oxygen to form an inactive molecule called oxyluciferin. The wavelength of light given off is between 510 and 670 nm (pale reddish green color). Amount of light emitted in this reaction is directly proportional to the ATP concentration. Measurements of ATP can be used to estimate microbial biomass

4

Diversity of Uncultivable Microorganism - Metagenomic Approach

Metagenomics and its application in soil microbial community studies

The use of traditional microbiological culturing methods for the study of microbes has had limited success. It has been estimated that 99% of the microbes cannot be cultivated easily. Different problems have been faced by researchers during attempts to culture some microbes. Over the past decade, "metagenomics," which is culture-independent genomics analysis of microbes, has been developed to overcome these difficulties. Soil is considered to be a complex environment, which is a major reservoir of microbial genetic diversity. Metagenomic approaches promise increased accessibility of the genetic resources contained in the soil.

Metagenomics

Metagenomics was coined by Jo Handelsman in 2004. metagenomics is a new field of research that allows the study of genomes recovered from environmental samples. Metagenomics can be defined as "the application of modern genomics techniques to the study of communities of microbial organisms directly in their natural environments. The metagenomic methodology has been developed as an effective tool for the discovery of new natural products and microbial functions.

Applications of metagenomics

It provides the capacity to effectively characterize the genetic diversity present in water, soil and rumen source samples regardless of the availability of laboratory culturing techniques. Metagenomics is a new and exciting field of molecular biology that is likely to grow into a standard technique for understanding biological diversity.

Metagenomics technologies

Metagenomics is a means of systematically investigating, classifying and manipulating the entire genetic material isolated from environmental samples. The process involves isolating DNA from environmental sources and cloning it into vectors that replicate in cultured organisms.

Metagenomics process which relies on the efficiency of three main steps, *i.e.*

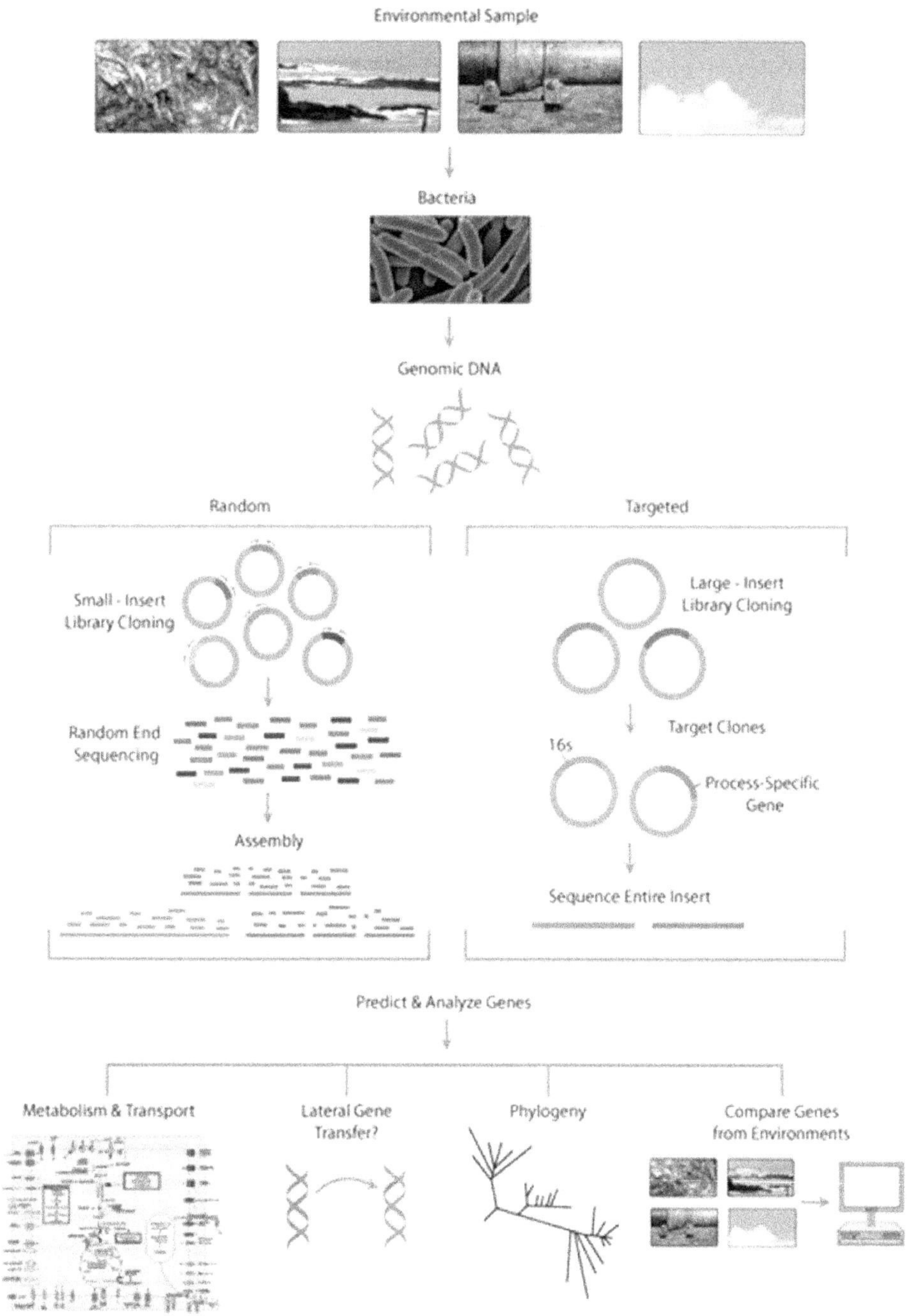

Steps involved in a metagenomics

1. Sampling and nucleic acids extraction
2. Library construction.
3. Analysis of metagenomic libraries

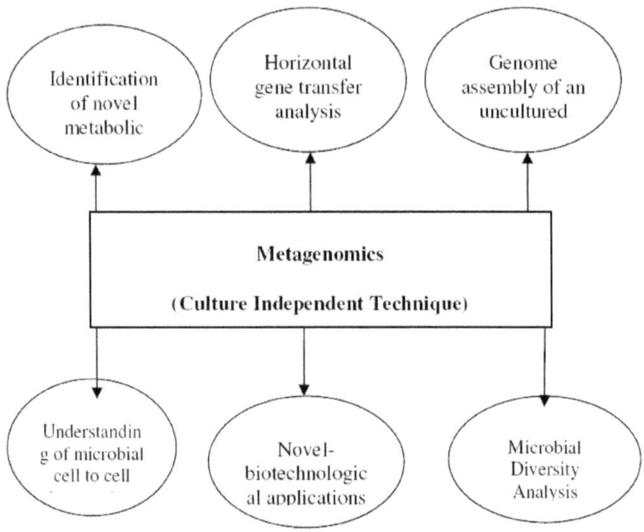

Applications of metagenomics.

Application of soil metagenomics

The development and application of metagenomics has enabled access to the uncultivated soil microbial community, availing a rich source of novel and useful biomolecules. Some examples of application of soil metagenomics are:

Target gene/natural product	Souce	Vector used for library construction	Host strain
Lipase, Esterase	Meadow Soil	PBluescript SK	*E. coli* DH5
Antibacterial, Hemolytic activities,Lipase,amylase,	North American Soil	PBeloBAC11	*E. coli* DH10B
nuclease 4-hydroxybutyrate dehydrogenase	Soil	PBluescript SK	*E. Coil* DH5
H+ antiporters	Soil	PBluescript SK	*E. coil* KNabc
Unique lipolytic activity	Forest Topsoil	Fosmid	*E.coil*
Novel biocatalysts	Unplanted field soil	Cosmid	*E.coil*
Alcohol oxidoreductase	Meadow, sugar beet field, cropland soil	PBluescript SK	*E.coil* DH5 alpha
Carbonyls formation	Soil	n.r	*E.coil*

Target gene/natural product	Souce	Vector used for library construction	Host strain
Coenzyme B (12)-dependent glycerol and diol dehydratases	Soil	PBluescript SK	*E.coil* DH5 alpha
esterase	Soil	pCCIOFS	*E.coil* EP-1300-TI
Lipase	Soil	pEpiFOS-5	*E.coil* EP1-100
Amidase	Soil and Enrichment Culture	PZero-2	*E.coil* TOP10
Amylase	Soil	Puc19	*E.coil* DH5 alpha
Polyketide synthase	Soil	Cosmid	*E.coil*
Aminoglycoside and tetracycline antibiotic	Soil	Pjn105	DH10B
resistance		pCF430	DH5 Alpha
Deoxybiolacein and broad spectrum antibiotic violacein	Soil	Cosmid	*E.coil*

Therefore, culture-independent methods are essential to understand the genetic diversity, population structure, and ecological roles of the majority of microorganisms. Metagenomics has potential to answer fundamental questions in microbial ecology, especially of soil ecosystems which contain the greatest diversity of microorganisms on earth, with 5,000–10,000 species of microorganism per gram of soil

More than 99% of bacteria from environmental samples remain 'unculturable' in the laboratory. Many of these 'unculturable' bacteria represent new phylotypes, families and divisions in domain bacteria and archaea. 'Unculturable' bacterial diversity presents a vast gene pool for biotechnological exploitation and poses a major challenge for microbiologists to understand their phylogenetic relationship and ecological significance. New culture techniques along with culture-independent methods, like PCR amplification from microbial community DNA (metagenome) and functional or sequence based screening of metagenomic DNA libraries are proving useful for exploitation of 'unculturable' bacteria. Environmental genome sequencing efforts are aiding better understanding of the community structure and their physiological importance in the ecosystem.

Definition

Metagenomics is the study of the collective genomes of the members of a microbial community. It involves cloning and analyzing the genomes without culturing the organisms in the community, thereby offering the opportunity to describe the planet's diverse microbial inhabitants, many of which cannot yet be cultured.

DNA is extracted directly from the community, cloned into a surrogate host, and then studied by sequencing or screening for expression of activities of interest. Many microbial communities have been tapped for metagenomic analyses. Following construction of a metagenomic library, two approaches can be taken to access the genomic information. Functional metagenomics requires that the host bacterium can express the recombinant DNA in either screens for active enzymes or antibiotic production or selections for growth under growth suppressive conditions (*e.g.*, nutrient deficiency or presence of an antibiotic). In sequence-based metagenomics, cloned DNA is randomly sequenced using vector-based primers or a specific gene is sought using complimentary oligonucleotides (oligos) to hybridize to arrayed metagenomic clones.

Culture-independent approach to assess the 'unculturable' bacterial diversity for novel biocatalysts and bioactive compounds

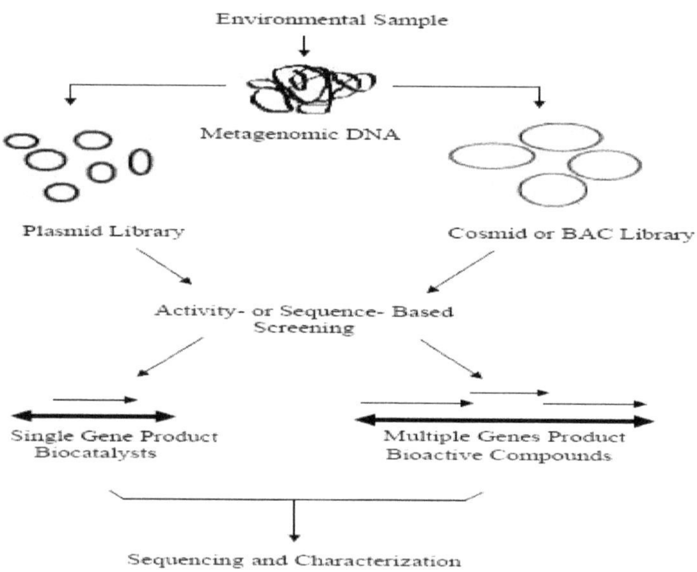

Strategies for the Metagenomic Analysis ofEnvironmental Microbial Communities

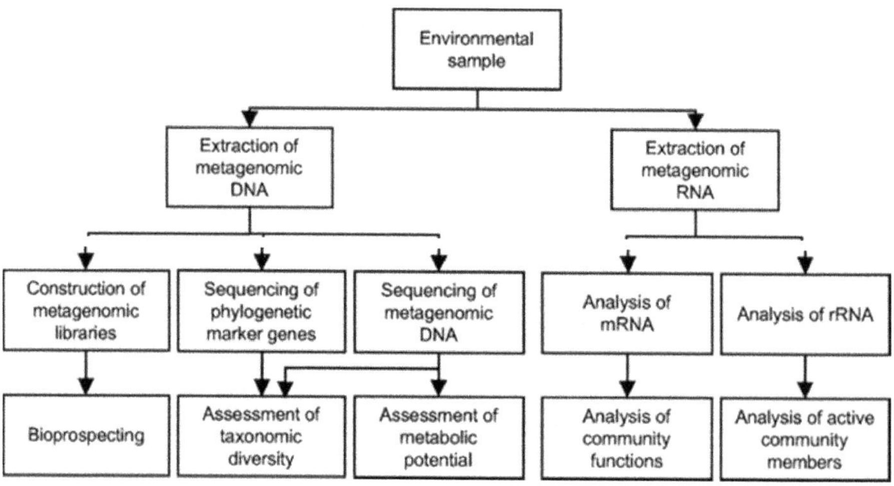

Metagenomics

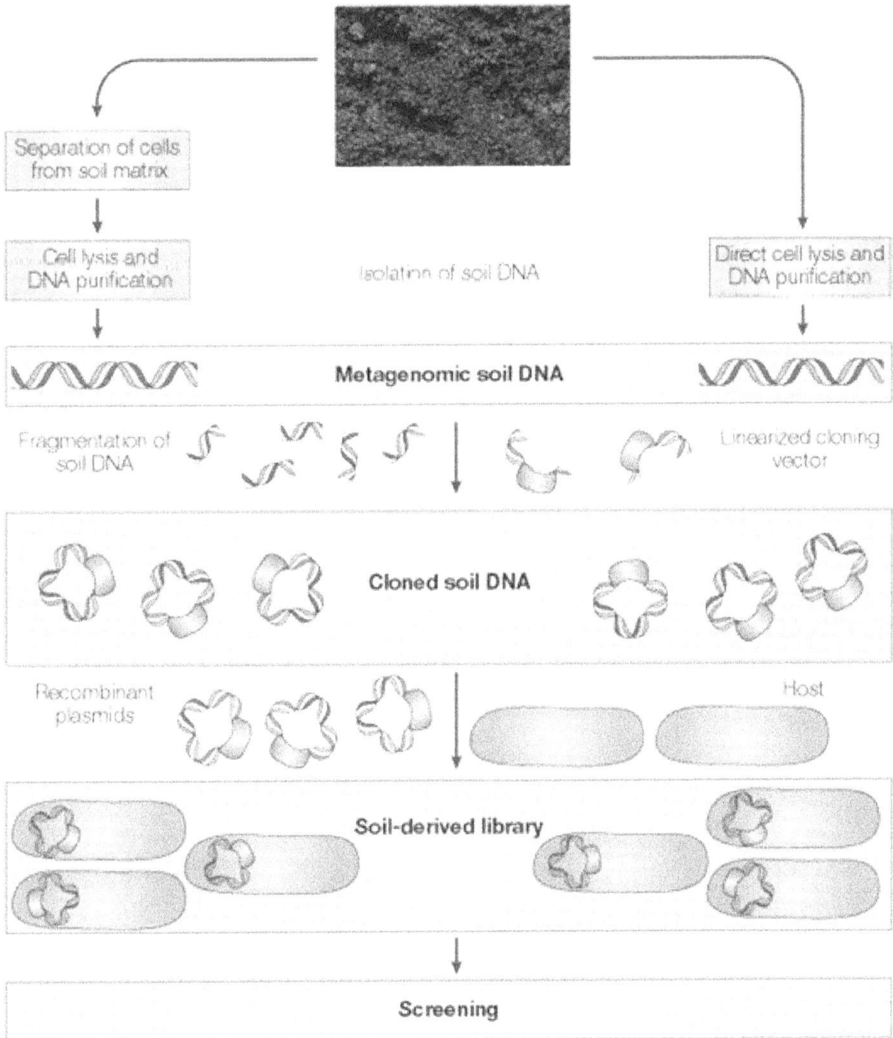

5

Rhizosphere and its Importance

Rhizosphere

The term Rhizosphere was introduced by the German scientist, Hiltner in 1904. It is region of the soil, which is subjected to the influence of plant roots. It is the unique environment under the influence of plant roots. The roots of plants are not only involved in the uptake of mineral nutrients and water for plant growth, but they also release a wide range of organic compounds in the surrounding soil.

Consequently, the area of soil in contact with the plant root, termed the rhizosphere, is a site of intense microbial activity. Not surprisingly, many microorganisms are present at higher numbers on the surface of plant roots and in the rhizosphere than in soil not influenced by the presence of roots. Hence rhizosphere is characterized by greater microbiological activity than the soil away from plant roots.

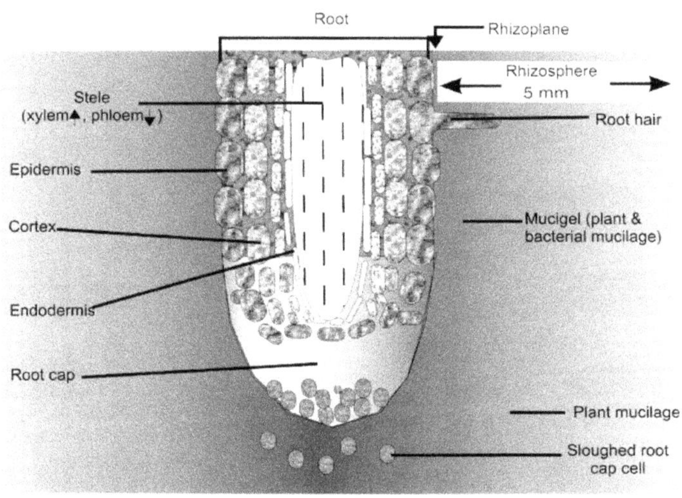

The term rhizosphere is often divided into two general areas, the inner rhizosphere (very root surface) and the outer rhizosphere (immediate adjacent soil). The microbial numbers are larger in the inner zone, where the bio chemical interactions between microorganism and roots are most pronounced.

The rhizosphere is a key soil habitat, where the numerous interactions taking place between plant root and soil microorganisms will determine growth conditions for both the plant and the microorganisms in the rhizosphere. Indeed, the rhizosphere microbiota exerts an important influence on roots and on growth of the plant. Consequently, the rhizosphere represents an important soil habitat where introduction of beneficial microorganisms (*e.g.* biofertilizers, phytostimulators, and biopesticides) as inoculants can result in significant improvements in crop yield and /or crop quality.

Rhizosphere Concept

The root system of higher plants is associated not only with soil environment composed of inorganic and organic matter, but also with a vast community of metabolically active microorganisms. As living plants create a unique habitat around the roots, the microbial population on and around the roots is considerably higher than that of root free soil environment and the differences may be both quantitative and qualitative.

1. **Rhizosphere:** It is the zone/region of soil immediately surrounding the plant roots together with root surfaces, or it is the region where soil and plant roots make contact, or it is the soil region subjected to influence of plant roots and characterized by increased microbial.

2. **Rhizoplane:** Root surface along with the closely adhering soil particles is termed as rhizoplane.

Rhizosphere effect

It indicates the over all influence of plant roots on soil microorganisms. R: S ratio is used to find out the degree or extend of influence of the plant roots on soil microorganisms.

R:S ratio is defined as the ratio of microbial numbers per unit weight of rhizosphere (R), to the numbers in a unit weight of the adjacent non-rhizosphere soil (S). The rhizosphere effect is consistently greater for bacteria than for other microbes. The short, gram-negative rods, which almost invariably make up to a larger percent of the rhizosphere. On a generic basis, root effect appears to be more pronounced on Pseudomonas, Flavobacterium, Alcaligenes,Arthrobacter, Brevibacterium, Cornebacterium, Micrococcus, Xanthomonas, and Bacillus. Anaerobic bacteria are affected by the root; this may be attributed to the reduced O_2 tension resulting from root and microbial respiration. Bacterial density in the rhizosphere soil is 10^9 per gram and it may cover 4-10 % of the root area. Fast growers and metabolic active population are dominating in the rhizosphere region. This suggests that the rhizosphere flora has a greater ability to effect rapid biochemical changes than the organisms of fallow land. Selective influence of he root system is exhibited in few cases like preferential stimulation of G-ve, non spore forming short rods, vitamin requiring bacteria, abundance of N_2 fixing and P solubilizing organisms *etc.*, The spectrum of genera varies with plant species, age and kind of the soil. Actinomycetes and algae are not significantly benefited by their proximity to roots.

Root exudates

One of the most important factors responsible for rhizosphere effect is the great variety of organic substances available at the root region by the way of exudates from roots, which directly or indirectly influence the quality and quantity of the micro organism in the root region. The substances exuded by plant roots include amino acids (asparagine,alanine, leucine, serine, glycine, lysin, prolein, trytophane), sugars (glucose, fructose, sucrose, xylose, arabinose, oligosaccharides), organic acids (tartaric, oxalic, malic, acetic, propionic, succinic, fumaric, butric), growth factors (biotin, thiamine, niacin, inositol, pyridoxin, p- amino benzoic acid), nucleotides, flavanones, enzymes and many other unidentified substances.

The nature and amount of substances exuded are dependent on the species of plant, age and environmental conditions under they grow. The products of photosynthesis are translocated to the root system in less than 12 hours and indicating the influence of metabolism of plants determining the rhizosphere effect. The root cap and ares of active growth are primary regions of root exudation.

Influence of the rhizosphere microflora on plant growth

The rhizosphere community has either a favourable or detrimental influence of plant development. Because the microflora is so intimately related with the root system, partially covering its surface, any beneficial or toxic substance produced can cause an immediate response. Vast microorganisms demand a variety of anions and cations for their own development and as a result immobilization take place. Aerobic organisms remove O_2 from the environment and add CO_2 and due to lowering of O_2 or increasing of CO_2 tension, it may reduce root elongation, development and reduce the water and nutrient uptake. It may however favour plant development by producing growth stimulating substances, forming of stable soil structure, releasing elements in organic forms through the mineralization of organic complexes, and by entering into symbiotic root associations.

Influence of the Plant on microflora

The microflora is affected in many ways by the growing plant. The important contribution is the provision of excretion products and sloughed off tissues to serve as energy, carbon, nitrogen or growth factors. Due to the large quantities of CO_2 liberated by the root zone during respiration, the carbonic acid is formed. It solubilizes the unavailable elements to make available and that will be assimilated by the microflora.

Historical Background

Term "Rhizosphere" was introduced for the first time by the German scientist **Hiltner(1904)** to denote that region of soil which is subjected to the influence of plant roots. The concept of "Rhizosphere Phenomenon" which shows the mutual interaction of roots and microorganisms was came into existence with the work of **Starkey et al (1929), Clark (1939) and Rauath and Katznelson (1957).**

N. V. Krassinikov (1934)found that free\ living nitrogen fixing bacteria, *Azotobacter*were unable to grow in the wheat rhizosphere.

Starkey (1938) examined the rhizosphere region of some plant species and demonstrated the effect of root exudates on the predominance of bacterial population in particular and other soil microorganisms in general in the rhizosphere region. Thus, he put forth the concept of **"Rhizosphere effect / phenomenon"** for the first time.

F E Clark (1949) introduced / coined the term **"Rhizoplane"** to denote the root surface together with the closely adhering soil particles.

R. I. Perotti (1925) suggested the boundaries of the rhizosphere region and showed that it was bounded on one side by the general soil region (called as Edaphosphere) and on the other side by the root tissues (called Histosphere).

G. Graf and S. Poschenrieder (1930) divided the rhizosphere region into two general areas *i.e.* outer rhizosphere and inner rhizosphere for the purpose of describing the same site of microbial action.

H. Katznelson (1946) suggested the **R:S ratio** *i.e.* the ratio between the microbial population in the rhizosphere (R) and in the soil (S) to find out the degree or extent of plant roots effect on soil microorganisms. R: S ratio gives a good picture of the relative stimulation of the microorganisms in the rhizosphere of different plant species.

R: S ratio is defined as the ratio of microbial population per unit weight of rhizosphere soil (R), to the microbial population per unit weight of the adjacent non-rhizosphere soil (S)

A. G. Lochhead and H. Katznelson (1940) examined in detail the qualitative differences between the microflora of the rhizosphere and microflora of the non-rhizosphere region and reported that gram-negative, rod shaped and non-spore forming bacteria are abundant in the rhizosphere than in the non-rhizosphere soil

C. Thom and H. Humfeld (1932) found that corn roots in acidic soils yielded predominantly *Trichoderma* while roots from alkaline soils mainly contained *Penicillium.*

M J. Timonin (1940) reported some differences in the fungal types and population in the rhizosphere of cereals and legumes. R: S ratio of fungal population was believed to be narrow in most of the plant species, usually not exceeding 10.

E. A. Peterson and others (1958) reported that the plant age and soil type influence the nature of fungal flora in the rhizosphere, and the number of fungal population gradually increases with the age of plant.

M. Adati (1932) studied many crops and found that though actinomycetes were relatively less stimulated than bacteria, but in some cases the R: S ratio of actinomycetes was as high as 62.

R. Venkatesan and G. Rangaswami (1965) studied the rhizosphere effect in rice plant on bacteria, actinomycetes and fungi and reported that (i) for actinornycetes R: S was more (ranging from 0 to 25) depending on the age of plant roots and the dominant genera reported were Nocardia, (ii) R:S ratio reduced with the depth of soil.

E. A. Gonsalves and V. S. Yalavigi (1960) reported the presence of greater number of algae in the rhizosphere

J. W. Rouatt et al reported positive rhizosphere effect on protozoa, but a negative effect on algae in wheat plants.

Microorganisms in the Rhizosphere and Rhizosphere Effect

The rhizosphere region is a highly favorable habitat for the proliferation, activity and metabolism of numerous microorganisms. The rhizosphere microflora can be enumerated intensively by microscopic, cultural and biochemical techniques. Microscopic techniques reveal the types of organisms present and their physical association with the outer root tissue surface / root hairs. The cultural technique most commonly followed is "serial dilution and plate count method" which reveal the quantitative and qualitative population of microflora. At the same time, a cultural method shows the selective enhancement of certain categories of bacteria. The biochemical techniques used are designed to measure a specific change brought about by the plant or by the microflora. The rhizosphere effect on most commonly found microorganisms *viz.* bacteria, actinomycetes, fungi, algae and protozoa is being discussed herewith in the following paragraphs.

A. Bacteria

The greater rhizosphere effect is observed with bacteria (R: S values ranging from 10-20 or more) than with actinomycetes and fungi. Gram-negative, rod shaped, non-sporulating bacteria which respond to root exudates are predominant in the rhizosphere *(Pseudomonas, Agrobacterium)*. While Gram-positive, rods, Cocci and aerobic spore forming *(Bacillus, Clostridium)* are comparatively rare in the rhizosphere. The most common genera of bacteria are: *Pseudomonas, Arthrobacter, Agrobacterium, Alcaligenes, Azotobacter, Mycobacterium, Flavobacter, Cellulomonas, Micrococcus* and others have been reported to be either abundant or sparse in the rhizosphere. From the agronomic point of view, the abundance of nitrogen fixing and phosphate solubilizing bacteria in the rhizosphere assumes a great importance. The aerobic bacteria are relatively less in the rhizosphere because of the reduced oxygen levels due to root respiration. The bacterial population in the rhizosphere is enormous in the ranging form 10^8 to 10^9 per gram of rhizosphere soil. They cover about 4-10% of the total root area occurring profusely on the root hair region and rarely in the root tips. There is predominance of amino acids and growth factors required by bacteria, are readily provided by the root exudates in the region of rhizosphere.

B. Fungi

In contrast to their effects on bacteria, plant roots do not alter / enhance the total count of fungi in the rhizosphere. However, rhizosphere effect is selective and significant on specific fungal genera *(Fusarium, Verticillium, Aspergillus* and *Penicillium)* which are stimulated. The R:S ratio of fungal population is believed to be narrow in most of the plants, usually not exceeding to 10. The soil / serial dilution and plating technique used for the enumeration of rhizosphere fungi may often give erratic results as most of the spore formers produce abundant colonies in culture media giving a wrong picture / estimate *(eg Aspergilli* and *Penicillia).* In fact the mycelial forms are more dominant in the field. The zoospore / forming lower fungi such as *Phytophthora, Pythium, Aphanomyces* are strongly attracted to the roots in response to particular chemical compounds excreted by the roots and cause diseases under favorable conditions. Several fungi *eg Gibberella* and *fujikurio* produces *phytohormones* and influence the plant growth.

C. Actinomycetes, Protozoa and Algae

Stimulation of actinomycetes in the rhizosphere has not been studied in much detail so far. It is generally understood that the actinomycetes are less stimulated in the rhizosphere than bacteria. However, when antagonistic actinomycetes increase in number they suppress bacteria. Actinomycetes may also increase in number when antibacterial agents are sprayed on the crop. Among the actinomycete, the phosphate solublizers (eg. *Nocardia, Streptomyces)* have a dominant role to play.

As rule actinomycetes, protozoa and algae are not significantly influenced by their proximity to the plant roots and their R: S ratios rarely exceed 2 to 3: 1 and around roots of plants, R: S ratio for these microorganisms may go to high. Because of large bacterial community, an increase in the number or activity of protozoa is expected in the rhizosphere. Flagellates and amoebae are dominant and ciliates are rare in the region.

Factors affecting microbial flora of the Rhizosphere / Rhizosphere Effect

The most important factors which affect / influence the microbial flora of the rhizosphere or rhizosphere effect are: soil type & its moisture, soil amendments, soil PH, proximity of root with soil, plant species, and age of plant and root exudates.

A. Soil type and its moisture: In general, microbial activity and population is high in the rhizosphere region of the plants grown in sandy soils and least in the high humus soils, and rhizosphere organisms are more when the soil moisture is low. Thus, the rhizosphere effect is more in the sandy soils with low moisture content.

B. Soil amendments and fertilizers: Crop residues, animal manure and chemical fertilizers applied to the soil cause no appreciable effect on the quantitative or qualitative differences in the microflora of rhizosphere. In general, the character of vegetation is more important than the fertility level of the soil.

C. Soil PH/ Rhizosphere PH: Respiration by the rhizosphere microflora may lead to the change in soil rhizosphere PH. If the activity and population of the rhizosphere microflora is more, then the PH of rhizosphere region is lower than that of surrounding soil or non-rhizosphere soil. Rhizosphere effect for bacteria and protozoa is more in slightly alkaline soil and for that of fungi is more in acidic soils.

D. Proximity of root with Soil: Soil samples taken progressively closer to the root system have increasingly greater population of bacteria, and actinomycetes and decreases with the distance and depth from the root system. Rhizosphere effect decline sharply with increasing distance between plant root and soil.

E. Plant Species: Different plant species inhabit often some what variable microflora in the rhizosphere region. The qualitative and quantitative differences are attributed to variations in the rooting habits, tissue composition and excretion products. In general, legumes show / produce a more pronounced rhizosphere effect than grasses or cereals. Biennials, due to their long growth period exert more prolonged stimulation on rhizosphere effect than annuals.

F. Age of Plant: The age of plant also alter the rhizosphere microflora and the stage of plant maturity controls the magnitude of rhizosphere effect and degree of response to specific microorganisms. The rhizosphere microflora increases in number with the age of the plant and reaching at peak during flowering which is the most active period of plant growth and metabolism. Hence, the rhizosphere effect was found to be more at the time of flowering than in the seedling or full maturity stage of the plants. The fungal flora (especially, Cellulolytic and Amylolytic) of the rhizosphere usually increases even after fruiting and the onset of senescence due to accumulation of moribund tissue and sloughed off root parts / tissues: whereas, bacterial flora of the rhizosphere decreases after the flowering period and fruit setting.

G. Root / exudates /excretion: One of the most important factors responsible for rhizosphere effect is the availability of a great variety of organic substances at the root region by way of root exudates/excretions. The quantitative and qualitative differences in the microflora of the rhizosphere from that of general soil are mainly due to influences of root exudates. The spectrum of chemical composition root exudates varies widely, and hence their influence on the microflora also varies widely.

Root exudates are composed of the chemical substances like

Sr. No	Root Executes	Chemical Substances
1	Amino Acids	All naturally occurring amino acids.
2	Organic acids	Acetic, butyric, citric, fumaric, lactic, malic, propionic, succinic *etc.*
3	Carbohydrates / sugars	Arabinose, fructose, galactose, glucose, maltose, mannose, oligosaccharides, raffinose, ribose, sucrose, xylose *etc.*
4	Nucleic acid derivatives	Adenine, cystidine, guanine, undine
5	Growth factors (phytohormones)	Biotin, choline, inositol, pyridoxine *etc*
6	Vitamins	Thiamine, nicotinic acid, biotin *etc*
7	Enzymes	Amylase, invertase, protease, phosphatase *etc.*
8	Other compounds	Auxins, glutamine, glycosides, hydrocyanic acid peptides, Uv-absorbing compounds, nematode attracting factors, spore germination stimulators, spore inhibitors *etc.*

The nature and amount of chemical substances thus exuded are dependent on the species of plant, plant age, inorganic nutrients, and temperature, light intercity, O2 / CO_2 level, root injury *etc*. Another source of nutrients for the microorganisms in the rhizosphere region is the sloughed off root epidermis which exert selective stimulation effect on some specific groups of microorganisms. For instance, glucose and amino acids in the exudates readily attract Gram-negative rods which predominantly colonize the roots. Sugars and amino acids in the root exudates stimulate the germination of chlamydospores and other resting spores of fungi; stimulation effect of root exudates on plant pathogenic fungi, nematodes is also well known.

Alterations in Rhizosphere Microflora

Foliar application of various chemicals leads to alterations in the rhizosphere microflora by changing the pattern of root exudates. The pattern of the rhizosphere microflora *i.e.* numbers and species composition can be changed / altered by various factors, such as: (i) Soil amendments, (ii) Foliar application of fertilizers / nutrients, fungicides, insecticides and hormones and (iii) Bacterization / microbial seed inoculants.

A. Soil amendments

Soil amendments with inorganic and organic fertilizers can alter the rhizosphere microflora and an understanding of the type of changes in the microflora can be useful in the indirect control of pathogens. Dwivedi and Chaube (1985) showed that amendment of soil with neem-cake can stimulate the activity of actinomycetes which results into the reduction of propagules of *Macrophomina phaseolina.* It is also known to control phytopathogenic nematodes in soil by stimulating nematode trapping fungi. Amendment of soil with castor and bean leaves stimulate the activity of *Trichoderma viride* and *Penicillium* in the rhizosphere leading to the control of *Sclerotium rolfsii.*

B. Foliar application of fertilizers and agrochemicals

Translocation of photosynthete from leaves to roots takes place as a part of the normal metabolic activity in plants. Therefore, organic substances, including plant protection chemicals (fungicides, insecticides), growth regulators and plant nutrients applied to foliage / leaves get absorbed into the leaf tissue and further get translocated to roots along with photosynthates. Many workers have reported that foliar application with various chemicals cause marked alterations in the number and kind / qualities of microorganisms in the rhizosphere of several cereals and leguminous crop plants. Thus, such an approach can be used as a new tool in the biological control of root diseases, stimulation of activity of nitrogen-fixing bacteria and other beneficial microorganisms in the soil.

C. Seed treatment with bio inoculants

Bio inoculants such as *Azotobacter, Beijerinckia, Azospirillum, Rhizobium or* P -solubilizing microorganisms (*e.g. Bcillus, plymyxa, Azotobacter croococcum, Aspergillus niger, Penicillium digitatum etc.*) When applied to the seed / soil helps in the establishment of beneficial microorganisms in the rhizosphere region which will further benefit in plant growth, encourage inhibition of plant pathogenic organisms in the root vicinity and enrich the soil with added microbial bio-mass.

Associative and Antagonistic activities in the Rhizosphere

In natural environments (*e.g.* Soil, Air, Water *etc.*) a number of relationships exist between individual microbes, microbial species and between individual cells.

The composition of microflora of any habitat (soil / rhizosphere) is governed by the biological equilibrium created by the associations and interactions of all individuals found in the community. In soil and rhizosphere region, many microorganisms live in close proximity and their interactions with each other may be associative or antagonistic.

A. Associative interactions / activities in rhizosphere: The dependence of one microorganism upon another for extra-cellular products (*e.g.* amino acids & growth promoting substances) can be regarded as an associative activity / effect in rhizosphere. There is an increase in the exudation of amino acids, organic acids and monosaccharide by plant roots in the presence of microorganisms. Gibberellins and gibberellin- like substances are known to be produced by bacterial genera *viz Azotobacter, Arthrobacter, Pseudomonas,* and *Agrobacterium* which are commonly found in the rhizosphere. Microorganisms also influence root hair development, mucilage secretion and lateral root development. Fungi inhabiting the root surface facilitate the absorption of nutrient by the roots.

Mycorrhiza is one of the best known associative / symbiotic interactions which exist between the roots of higher plants and fungi. This mycorrhizal association has been found to improve plant growth through better uptake of phosphorus and zinc from soil, suppression of root pathogenic fungi and nematodes. Another example is association between the bacterium *Rhizobium* and roots of legumes and *Azospirillum* with cereal crops (wheat, rye, bajara, maize *etc*).

B. Antagonistic interactions / activities in rhizosphere: The biochemical qualities of root exudates and the presence of antagonistic microorganisms, plays important role in encouraging or inhibiting the soil borne plant pathogens in the rhizosphere region. Several mutualistic, communalistic, competitive and antagonistic interactions exist in the rhizosphere. The number and qualities of antagonistic microorganisms in the rhizosphere could be increased through artificial means such as fertilizer application, organic amendments, foliar spraying of chemicals *etc*.

Antagonistic microorganisms in the rhizosphere play an important role in controlling some of the soil borne plant pathogens. Stanier et al (1966) discovered the bacterial strain *Pseudomonas fluorescens* and the fluorescent pigments of this species in biological control of root pathogens. Strains of *P. fluorescence* are collectively called as **"Fluorescent Pseudomonads"**. They produce variety of biologically active compounds such as plant growth substances, cyanides, antibiotics and iron chelating substances called **"Siderophores"** Rovira and Campbell (1975), showed that bacterial strains of *P fluorescens* could lyse the hyphae of *Gaumannomyces graminis var. Tritici,* the causative agent of take-all disease of wheat. Fluorescent pseudomonads *(P. fluorescens, P. putida)* are known to produce iron chelating substances called Siderophores. These are low molecular weight, extra cellular, iron-binding agents produced by pseudomonads in response to low iron stress or when Fe3 is in short supply. Thus, iron stress triggers the formation of iron-binding ligands called siderophores. Siderophores contains the pigments Pyovirdin (Fluorescent) and Pyocyanin (non-Fluorescent) having iron chelating properties. Another pigment "Pseudobactin" is a fluorescent chelator of iron which is known to promote plant growth and inhibition of pathogenic bacteria in the rhizosphere.

An antibiotic called "Pyrrolnitrin" reduces damping-off disease in cotton caused by *Rhizoctonia solani.* Several species of *Bacillus* are known to cause mycolysis in the rhizosphere. *eg. Fusarium oxysporum* hyphae are known to undergo lysis in soil due to these bacterial metabolites.

The successful antagonists among fungi are *Trichoderma* sp (T. viride and *T. harzianum, T. hamatum)* and *Gliocladium virens* which parasitize, lyse or kill the phytopathogenic fungi in the soil. Antifungal and antibacterial actinomycetes in the rhizosphere play an important role in controlling pathogenic fungi and bacteria, for example *Micromonospora globosa* is a potent antagonist of *Fusarium udum* causing wilt of pigeon pea. Amoebae are also known to play an antagonistic role in controlling soil fungi, *e.g.* control of take-all disease of wheat caused by *Gaumannomyces graminis* through the use of Myxamoebae. There can also occur antagonisms between two fungi producing metabolite and interfering the growth of the other fungus as in case of *Peniophora* antagonizing *Heterobasidium.*

Rhizosphere in relation to Plant Pathogens

Plant root exudates influence pathogenic fungi, bacteria and nematodes in various ways. The effect may be in the form of attraction of fungal zoospores, or bacterial cells towards the roots; stimulation of germination of dormant spores and hatching of cysts of nematodes. Root exudates may contain inhibitory substances preventing the establishment of pathogens. The balance between the rhizosphere microflora and plant pathogens and soil microflora and plant pathogens is important in host-pathogenic relationship. In this context, the biochemical qualities of root exudates and the presence of antagonistic micro-organisms plays an important role in the proliferation and survival of root infecting pathogens in soil either through soil fungi stasis, inhibition or antibiosis of pathogens in the rhizosphere.

Some of the most common interactions between plant roots and plant pathogenic microorganisms in the rhizosphere are discussed herewith.

A. Zoospore attraction: Amino acids, organic acids and sugars in the root exudates stimulate the movement and attraction of zoospores towards root of the plants. For example attraction of zoospores has been reported in *Phytophthora citrophthora* (Citrus roots), *P. parasitica* (tobacco roots) and *Pythium aphanidermatum* (pea root).

B. Spore germination: The spores or conidia of many pathogenic fungi such as *Rhizoctonia, Fusarium, Sclerotium, Pythium, Phytophthora* etc. have been stimulated to germinate by the root exudates of susceptible cultivars of the host plants. There are some reports on the selective stimulation of *Fusarium, Pseudomonas* and root infecting nematodes in the rhizosphere region of the respective susceptible hosts. This stimulus to germination is especially important to those plant pathogens which are not vigorous competitors and remain in resting stage due to shortage of nutrients or fungistasis. As a rule, germination and subsequent hyphal development are promoted by non host species and also by both susceptible and resistant cultivars of the host plants. The quantity and quality of microorganisms present in the rhizosphere of disease resistant crop varieties are significantly different from those of susceptible varieties.

C. Changes in morphology and physiology of host plant: Changes in the physiology and morphology of host plant influence the rhizosphere microflora through root exudations. Hence, significant changes in the rhizosphere microflora of diseased plants were reported which are attributed to the nature and severity of the disease. Systemic virus diseases cause marked changes in the plant morphology and physiology to drastically alter the rhizosphere microflora.

D. Increase in antagonists activity: Root exudates provide a food base for the growth of antagonistic organisms which plays an important role in controlling / suppressing some of the soil borne plant pathogens. Generally, rhizosphere of the resistant plant varieties harboure moer number of *Streptomyces* and *Trichoderma* than that of susceptible varieties. For example in the rhizosphere of pigeon pea varieties resistant to *Fusarium udum,* the population of *Streptomyces* was found more which inhibited the growth of the pathogen. High density of *Trichoderma viride* in the rhizosphere of Tomato varieties resistant to *Verticillium* wilt has been reported with its ability to reduce the severity of wilt in susceptible plants.

E. Inhibition of pathogen: Root exudates containing toxic substances such as glycosides and hydrocyanic acid may inhibit the growth of pathogens in the rhizosphere. It has been reported that root exudates from resistant varieties of Flax (*e.g.* Bison) excrete a glucoside which on hydrolysis produces hydrocyanic acid that inhibits Fusarium oxysporum, the flax root pathogen. Exudates of resistant pea reduce the germination of spores of Fusarium oxysporum.

In this light, the rhizosphere may be considered as a microbiological buffer zone in which the microflora serves to protect the plants against the attack of the pathogens.

F. Attraction of bacteria and nematodes: Root exudates attracts phytopathogenic bacteria and fungi in the rhizosphere for example *Agrobacterium tumefaciens* have been reported to be attracted to the roots of the host plants like peas, maize, onion, tobacco, tomato and cucumber.

Host root exudates also influence phytopathogenic nematodes in two ways: (i) though stimulation of egg-hatching process and (ii) attraction of larvae towards plant roots.

6

Plant Growth Promoting Rhizobacteria (PGPR)

Rhizobacteria are root-colonizing bacteria that form symbiotic relationships with many plants. The name comes from the Greek rhiza, meaning root. Though parasitic varieties of rhizobacteria exist, the term usually refers to bacteria that form a relationship beneficial for both parties (mutualism). They are an important group of microorganisms used in biofertilizer. Biofertilization accounts for about 65% of the nitrogen supply to crops worldwide. Rhizobacteria are often referred to as plant growth-promoting rhizobacteria, or PGPRs. The term PGPRs was first used by Joseph W. Kloepper in the late 1970s and has become commonly used in scientific literature. PGPRs have different relationships with different species of host plants. The two major classes of relationships are rhizospheric and endophytic. Rhizospheric relationships consist of the PGPRs that colonize the surface of the root, or superficial intercellular spaces of the host plant, often forming root nodules. The dominant species found in the rhizosphere is a microbe from the genus Azospirillum. Endophytic relationships involve the PGPRs residing and growing within the host plant in the apoplastic space.

Nitrogen fixation

Nitrogen fixation is one of the most beneficial processes performed by rhizobacteria. Nitrogen is a vital nutrient to plants and gaseous nitrogen (N_2) is not available to them due to the high energy required to break the triple bonds between the two atoms. Rhizobacteria, through nitrogen fixation, are able to convert gaseous nitrogen (N_2) to ammonia (NH_3) making it an available nutrient to the host plant which can support and enhance plant growth. The host plant provides the bacteria with amino acids so they do not need to assimilate ammonia. The amino acids are then shuttled back to the plant with newly fixed nitrogen. Nitrogenase is an enzyme involved in nitrogen fixation and requires anaerobic conditions. Membranes within root nodules are able to provide these conditions. The rhizobacteria require oxygen to metabolize, so oxygen is provided by a hemoglobin protein called leghemoglobin which is produced within the nodules. Legumes are well-known nitrogen-fixing crops and have been used for centuries in crop rotation to maintain the health of the soil.

Symbiotic relationships

The symbiotic relationship between rhizobacteria and their host plants is not without costs. For the plant to be able to benefit from the added available nutrients provided by the rhizobacteria, it needs to provide a place and the proper conditions for the rhizobacteria to live. Creating and maintaining root nodules for rhizobacteria can cost between 12–25% of the plant's total photosynthetic output. Legumes are often able to colonize early successional environments due to the unavailability of nutrients. Once colonized, though, the rhizobacteria make the soil surrounding the plant more nutrient rich, which in turn can lead to competition with other plants. The symbiotic relationship, in short, can lead to increased competition.

PGPRs increase the availability of nutrients through the solubilization of unavailable forms of nutrients and by the production of siderophores which aids in the facilitating of nutrient transport. Phosphorus, a limiting nutrient for plant growth, can be plentiful in soil, but is most commonly found in insoluble forms. Organic acids and phosphotases released by rhizobacteria found in plant rhizospheres facilitate the conversion of insoluble forms of phosphorus to plant-available forms such as $H_2PO_4^-$. PGPR bacteria include Pseudomonas putida, Azospirillum fluorescens, and Azospirillum lipoferum and notable nitrogen-fixing bacteria associated with legumes includes Allorhizobium, Azorhizobium, Bradyrhizobium, and Rhizobium.

Though microbial inoculants can be beneficial for crops, they are not widely used in industrial agriculture, as large-scale application techniques have yet to become economically viable. A notable exception is the use of rhizobial inoculants for legumes such as peas. Inoculation with PGPRs ensures efficient nitrogen fixation, and they have been employed in North American agriculture for over 100 years.

Plant growth-promoting rhizobacteria

Plant growth-promoting rhizobacteria (PGPR) were first defined by Kloepper and Schroth to describe soil bacteria that colonize the roots of plants following inoculation onto seed and that enhance plant growth. The following are implicit in the colonization process: ability to survive inoculation onto seed, to multiply in the spermosphere (region surrounding the seed) in response to seed exudates, to attach to the root surface, and to colonize the developing root system. The ineffectiveness of PGPR in the field has often been attributed to their inability to colonize plant roots. A variety of bacterial traits and specific genes contribute to this process, but only a few have been identified. These include motility, chemotaxis to seed and root exudates, production of pili or fimbriae, production of specific cell surface components, ability to use specific components of root exudates, protein secretion, and quorum sensing. The generation of mutants altered in expression of these traits is aiding our understanding of the precise role each one plays in the colonization process.

Progress in the identification of new, previously uncharacterized genes is being made using nonbiased screening strategies that rely on gene fusion technologies. These strategies employ reporter transposons and in vitro expression technology (IVET) to detect genes expressed during colonization.

Using molecular markers such as green fluorescent protein or fluorescent antibodies, it is possible to monitor the location of individual rhizobacteria on the root using confocal laser scanning microscopy. This approach has also been combined with an rRNA-targeting probe to monitor the metabolic activity of a rhizobacterial strain in the rhizosphere and showed that bacteria located at the root tip were most active.

Mechanisms of action

PGPRs enhance plant growth by direct and indirect means, but the specific mechanisms involved have not all been well-characterized, Direct mechanisms of plant growth promotion by PGPRs can be demonstrated in the absence of plant pathogens or other rhizosphere microorganisms, while indirect mechanisms involve the ability of PGPRs to reduce the harmful effects of plant pathogens on crop yield. PGPRs have been reported to directly enhance plant growth by a variety of mechanisms: fixation of atmospheric nitrogen transferred to the plant, production of siderophores that chelate iron and make it available to the plant root, solubilization of minerals such as phosphorus, and synthesis of phytohormones. Direct enhancement of mineral uptake due to increases in specific ion fluxes at the root surface in the presence of PGPRs has also been reported. PGPR strains may use one or more of these mechanisms in the rhizosphere. Molecular approaches using microbial and plant mutants altered in their ability to synthesize or respond to specific phytohormones have increased understanding of the role of phytohormone synthesis as a direct mechanism of plant growth enhancement by PGPRs. PGPR that synthesize auxins and cytokinins or that interfere with plant ethylene synthesis have been identified.

Pathogenic roles

Studies conducted on sugar beet crops found that some root-colonizing bacteria were deleterious rhizobacteria (DRB). Sugar beet seeds inoculated with DRB had reduced germination rates, root lesions, reduced root elongation, root distortions, increased fungi infection, and decreased plant growth. In one trial the sugar beet yield was reduced by 48%.

Six strains of rhizobacteria have been identified as being DRB. The strains are in the genera Enterobacter, Klebsiella, Citrobacter, Flavobacterium, Achromobacter, and Arthrobacter. Due to a large number of taxonomic species yet to be described, complete characterization has not been possible as DRB are highly variable.

The presence of PGPRs has proven to reduce and inhibit the colonization of DRB on sugar beet roots. Plots inoculated with PGPRs and DRBs had an increase in production of 39% while plots only treated with DRBs had a reduction in production of 30%.

Biocontrol

Rhizobacteria are also able to control plant diseases that are caused by other bacteria and fungi. Disease is suppressed through induced systematic resistance and through the production of antifungal metabolites. Pseudomonas biocontrol strains have been genetically modified to improve plant growth and improve the disease resistance of agricultural crops. In agriculture, inoculant bacteria are often applied to the seed coat of seeds prior to being sown. Inoculated seeds are more likely to establish large enough rhizobacterial populations within the rhizosphere to produce notable beneficial effects on the crop.

The group of bacteria that colonize roots or rhizosphere soil and beneficial to crops are referred to as plant growth promoting rhizobacteria (PGPR). The PGPR inoculants currently commercialized that seem to promote growth through at least one mechanism; suppression of plant disease (termed Bioprotectants), improved nutrient acquisition (termed Biofertilizers), or phytohormone production (termed Biostimulants).

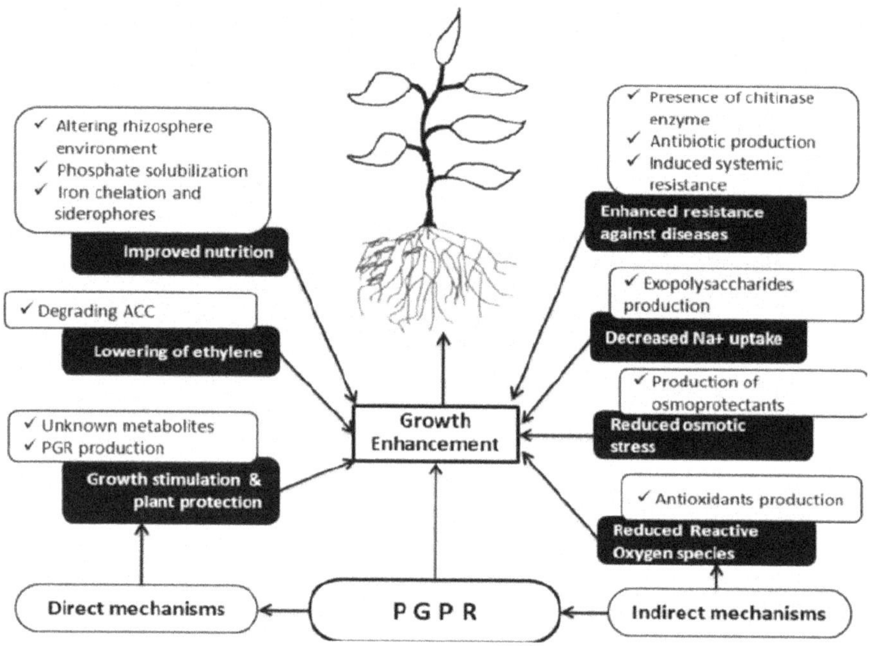

Species of Pseudomonas and Bacillus can produce as yet not well characterized phytohormones or growth regulators that cause crops to have greater amounts of fine roots which have the effect of increasing the absorptive surface of plant roots for uptake of water and nutrients. These PGPR are referred to as Biostimulants and the phytohormones they produce include indole-acetic acid, cytokinins, gibberellins and inhibitors of ethylene production.

Recent advances in molecular techniques also are encouraging in that tools are becoming available to determine the mechanism by which crop performance is improved using PGPR and track survival and activity of PGPR organisms in soil and roots. The science of PGPR is at the stage where genetically modified PGPR can be produced. PGPR with antibiotic, phytohormone and siderophore production can be made.

Despite of promising results, biofertilizers has not got widespread application in agriculture mainly because of the variable response of plant species or genotypes to inoculation depending on the bacterial strain used. Differential rhizosphere effect of crops in harbouring a target strain or even the modulation of the bacterial nitrogen fixing and phosphate solubilizing capacity by specific root exudates may account for the observed differences. On the other hand, good competitive ability and high saprophytic competence are the major factors determining the success of a bacterial strain as an inoculant.

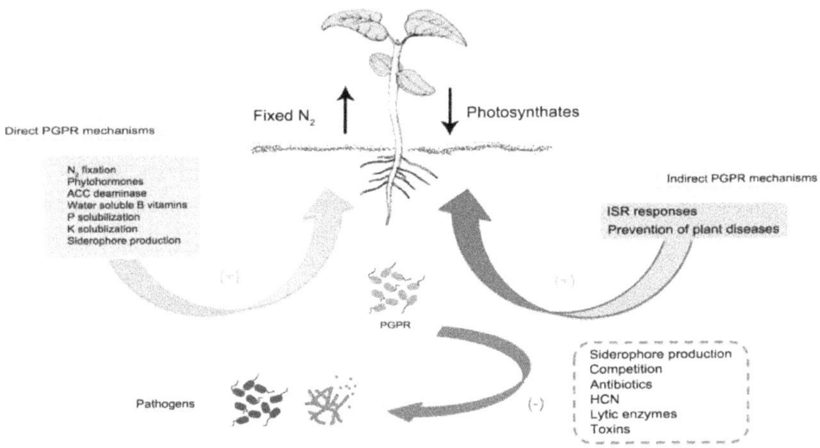

Studies to know the synergistic activities and persistence of specific microbial populations in complex environments, such as the rhizosphere, should be addressed in order to obtain efficient inoculants.

Phyllosphere Bacteria

Aerial plant surfaces represent the largest biological interface on Earth and provide essential services as sites of carbon dioxide fixation, molecular oxygen release, and primary biomass production. Rather than existing as axenic organisms, plants are colonized by microorganisms that affect both their health and growth.

For terrestrial plants, the phyllosphere represents the interface between the above-ground parts of plants and the air. Conservative estimates indicate that the roughly 1 billion square kilometers of worldwide leaf surfaces host more than 10^{26} bacteria, which are the most abundant colonizers of this habitat. The overall microbiota in this ecosystem is thus sufficiently large to have an impact on the global carbon and nitrogen cycles. Additionally, the phyllosphere inhabitants influence their hosts at the level of the individual plants. To a large extent, interest in phyllosphere microbiology has been driven by investigations on plant pathogens. Their spread, colonization, survival, and pathogenicity mechanisms have been the subject of numerous studies. Much less understood are nonpathogenic microorganisms that inhabit the phyllosphere. The composition of the phyllosphere microbiota has been analyzed in only a few studies by cultivation-independent methods; however, such methods are essential in light of the yet uncultivated majority of bacteria existing in nature, or more specifically on plant leaves. Not only their identity, but in particular the physiological properties of phyllosphere bacteria, their adaptations to the habitat, and their potential role (*e.g.*, with respect to modulating population sizes of pathogens) remain largely unknown. Current knowledge on the traits important in the phyllosphere is derived from relatively few studies on gene expression and stems mostly from model bacteria cultivated on host plants under controlled conditions.

However, under natural conditions, plants and their residing microorganisms are exposed to a host of diverse, highly variable environmental factors, including UV light, temperature, and water availability; moreover, individual microbes are subjected to competition with other microorganisms over resources, such as nutrients and space.

Toward a deeper understanding of phyllosphere microbiology, and in particular to learn more about the commensal majority of plant leaf colonizing bacteria, which may be of relevance for plant health and development, integrated approaches are needed. Here, Bacterial communities in the phyllosphere are thought to be limited by carbon availability, and it may be expected that access to carbon compounds on leaves is a major determinant of epiphytic colonization. There is evidence that small amounts of nutrients, such as simple sugars including glucose, fructose, and sucrose, leach from the interior of the plant.

The above-ground parts of plants are normally colonized by a variety of bacteria, yeasts, and fungi. While a few microbial species can be isolated from within plant tissues, many more are recovered from the surfaces of healthy plants. The aerial habitat colonized by these microbes is termed the phyllosphere, and the inhabitants are called epiphytes. While there has been some investigation of the colonists of buds and flowers, most work on phyllosphere microbiology has focused on leaves, a more dominant aerial plant structure. Bacteria are by far the most numerous colonists of leaves, often being found in numbers averaging 10^6 to 10^7 cells/cm2 (up to 10^8 cells/g) of leaf.

Compared to most other bacterial habitats, there has been relatively little examination of phyllosphere microbiology. This is somewhat surprising given the abundance of plants in the world and the roles of various phyllosphere bacteria in the important processes discussed below. Leaves constitute a very large microbial habitat.

It is estimated that the terrestrial leaf surface area that might be colonized by microbes is about 6.4×10^8 km^2. Given the large number of bacteria on leaves in temperate regions of the world and that population in tropical regions are probably even larger, the planetary phyllosphere bacterial population may be as large as 10^{26} cells. Clearly, in aggregate, these bacteria are sufficiently numerous to contribute in many processes of importance to global processes, as well as to the behavior of the individual plants on which they live. The microbial communities of leaves are diverse and include many different genera of bacteria, filamentous fungi, yeasts, algae, and, less frequently, protozoa and nematodes. Filamentous fungi are considered transient inhabitants of leaf surfaces, being present predominantly as spores, whereas rapidly sporulating species and yeasts colonize this habitat more actively.

Bacteria are by far the most abundant inhabitants of the phyllosphere. Epiphytic bacterial populations differ sharply in size among and within plants of the same species, as well as in close proximity, and over short time scales as well as over the growing season. These considerable variations in population sizes are caused in great part by the large fluctuations in the physical and nutritional conditions characteristic of the phyllosphere. Additionally, plant species appear to influence the microbial carrying capacity of the leaf, since the total number of culturable bacteria recovered from broad-leaf plants such as cucumber and beans was significantly greater than that recovered from grasses or waxy broad-leaf plants. Reflective of marked differences in the physicochemical environments of above-ground versus subterranean plant surfaces, the leaf bacterial flora differs substantially from that of roots. For example, pigmented bacteria, which are rarely found in the rhizosphere, dominate leaf surfaces, presumably because solar radiation influences the ecology of the phyllosphere. The differential composition of leaf and root bacterial communities is further evidenced by the failure of common root colonizers such as Rhizobium and Azospirillum to become established on leaves.

Studies of the composition of bacterial communities on leaves have been numerous but rather limited in scope. It is generally believed that populations of culturable aerobic bacteria on leaves are dominated by a few genera. A few exhaustive studies of the variations in the microbial community of leaves over multiple time and space scales have provided important detailed knowledge about the identity and the ecology of bacterial leaf inhabitants. Ercolani made an extensive inventory of culturable aerobic bacteria isolated from the surface of olive leaves over six growing seasons and reported distinct bacterial community structures on leaves of the same age at a given time of the growing season.

Most importantly, like Ercolani, they found distinct patterns of microbial colonization at different times of the year, with bacterial community diversity being lowest during the warmest and driest months of the season and highest during the cooler and rainy months. Coincidentally, in both of the above described studies, communities on young leaves were composed of a greater number of taxa than those of old leaves. Thus, specific natural environments of the phyllosphere apparently select for the presence of specific genotypes within the leaf bacterial community. This is further supported by the finding that the acquisition by Pseudomonas fluorescens of plasmids that are indigenous to the leaf microflora coincided with a specific maturation stage of the plant over two consecutive years. This indicated that traits carried on these plasmids conferred variable selective fitness to specific plasmid-bacterial host combinations during the growing season, possibly in response to changing conditions in the phyllosphere habitat.

The study of bacterial colonizers of leaves has been restricted mostly to aerobic culturable bacteria and also driven by the importance of investigating the ecology of plant-pathogenic bacteria because of their deleterious effect on plant productivity. Thus, the microbial ecology of the phyllosphere has been viewed mainly through the biology of gram-negative bacteria such as Pseudomonas syringae and Erwinia (Pantoea) spp., two of the most ubiquitous bacterial participants of phyllosphere communities. There is reason to believe, however, that the extreme fluctuations in

Soil and Applied Microbiology

The Author

Dr. V. Ponvizhi Ramya has completed her graduation from Tamil Nadu Agricultural University in 2009. She has done her *M.Sc.* and *Ph.D* programme in Agricultural Microbiology from the same University. She has received a National Award from ICAR and NAIP for her innovative idea in her Doctoral research. She has published many research articles in reputed journals. Currently she is working as Assistant Professor in Agricultural Microbiology at Thanthai Roever Institute of Agriculture and Rural Development, Chennai - Trichy Hwy, Tamil Nadu.

Soil and Applied Microbiology

V. Ponvizhi Ramya

2020

Associated Publishing Company®

A Division of

Astral International Pvt. Ltd.

New Delhi – 110 002

ISBN 9789390371730 (Int. Edition)

Published by : **Associated Publishing Company®**
A Division of
Astral International Pvt. Ltd.
– ISO 9001:2015 Certified Company –
4736/23, Ansari Road, Darya Ganj
New Delhi-110 002
Ph. 011-43549197, 23278134
E-mail: info@astralint.com
Website: www.astralint.com

Preface

It gives great pleasure and satisfaction to fulfill the long felt need of microbiologists and students for a comprehensive on Soil and Applied Microbiology. Experience gained from teaching agricultural microbiology encouragement gained from the university, faculty members and library facilities have made the dream come true. The modern metagenomic approach, PGPR and PPFM have been present in a readable form so that it will be student friendly. This book will definitely pave the way for the young aspiring microbiologist mind to take up the subject with confidence.

Several of my colleagues have contributed in bringing out this book. My family members has helped in preparing this script and I greatly indebted to all of them.

Dr. V. Ponvizhi Ramya

Unit - I: Soil Microbiology

1

Introduction to Soil Microbiology

Introduction

Soil represents a medium or substrate in which numerous microorganisms live and bring about a great variety of processes that are responsible for continuation of the cycle of life in nature. The numerous living forms which spend their life in soil, ranging from sub microscopic forms to the lower animal forms. With the growing recognition of the numerous processes carried out by the microorganisms in the soil there gradually emerged a branch of microbiology, which came to known as soil microbiology. It is a branch of soil science concerned with soil inhabiting microorganisms and their functions and activities.

Since soil microbiology concerns with soil microorganisms and their processes it is closely associated with soil biochemistry. Medical bacteriologists were interested in the soil as a medium for the growth and survival of disease producing organisms. Agricultural chemists are also interested in the soil processes that result from the activities of microorganisms. General bacteriologist, zoologist, botanist were interested in certain special group of organisms found in soil. Recently, soil microbiology has expanded to include the study of the role of soil microorganisms in genetic engineering, in the biological control of pests and diseases, the degradation of pollutants, production and destruction of radioactive gases and its transfer. Thus microbial participation in several important processes emphasizes that soil microbiology has become a global science.

Soil Microbiology

It is branch of science/microbiology which deals with study of soil microorganisms and their activities in the soil.

- Deals with the microscopic organisms of the soil
- Their population and activities
- Role of microorganisms in various transformations taking place in the soil and
- Their importance in plant nutrition and crop production

Soil

It is the outer, loose material of earth's surface which is distinctly different from the underlying bedrock and the region which support plant life. Agriculturally, soil is the region which supports the plant life by providing mechanical support and nutrients required for growth. From the microbiologist view point, soil is one of the most dynamic sites of biological interactions in the nature. It is the region where most of the physical, biological and biochemical reactions related to decomposition of organic weathering of parent rock take place.

Components of Soil

Soil is an admixture of five major components *viz.* organic mater, mineral matter, soil-air, soil water and soil microorganisms/living organisms. The amount/ proposition of these components varies with locality and climate.

1. **Mineral / Inorganic Matter:** It is derived from parent rocks/bed rocks through decomposition, disintegration and weathering process. Different types of inorganic compounds containing various minerals are present in soil. Amongst them the dominant minerals are Silicon, Aluminum and iron and others like Carbon, Calcium Potassium, Manganese, Sodium, Sulphur, Phosphorus *etc.* are in trace amount. The proportion of mineral matter in soil is slightly less than half of the total volume of the soil.
2. **Organic matter/components:** Derived from organic residues of plants and animals added in the soil. Organic matter serves not only as a source of food for microorganisms but also supplies energy for the vital processes of metabolism which are characteristics of all living organisms. Organic matter in the soil is the potential source of N, P and S for plant growth. Microbial decomposition of organic matter releases the unavailable nutrients in available from. The proportion of organic matter in the soil ranges from 3-6% of the total volume of soil.
3. **Soil Water:** The amount of water present in soil varies considerably. Soil water comes from rain, snow, dew or irrigation. Soil water serves as a solvent and carrier of nutrients for the plant growth. The microorganisms inhabiting in the soil also require water for their metabolic activities. Soil water thus, indirectly affects plant growth through its effects on soil and microorganisms. Percentage of soil-water is 25% total volume of soil.
4. **Soil air (Soil gases):** A part of the soil volume which is not occupied by soil particles *i.e.* pore spaces are filled partly with soil water and partly with soil air. These two components (water & air) together only accounts for approximately half the soil›s volume. Compared with atmospheric air, soil is lower in oxygen and higher in carbon dioxide, because CO_2 is continuous recycled by the microorganisms during the process of decomposition of organic matter. Soil air comes from external atmosphere and contains nitrogen, oxygen CO_2 and water vapour (CO_2> oxygen). CO_2 in soil air (0.3-1.0%) is more than atmospheric air (0.03%). Soil aeration plays important role in plant growth, microbial population, and microbial activities in the soil.

5. **Soil microorganisms:** Soil is an excellent culture media for the growth and development of various microorganisms. Soil is not an inert static material but a medium pulsating with life. Soil is now believed to be dynamic or living system.

Soil contains several distinct groups of microorganisms and amongst them bacteria, fungi, actinomycetes, algae, protozoa and viruses are the most important. But bacteria are more numerous than any other kinds of microorganisms. Microorganisms form a very small fraction of the soil mass and occupy a volume of less than one percent. In the upper layer of soil (top soil up to 10-30 cm depth *i.e.* Horizon A), the microbial population is very high which decreases with depth of soil. Each organisms or a group of organisms are responsible for a specific change / transformation in the soil. The final effect of various activities of microorganisms in the soil is to make the soil fit for the growth & development of higher plants. Living organisms present in the soil are grouped into two categories as follows.

1. Soil flora (micro flora) *e.g.* Bacteria, fungi, Actinomycetes, Algae and
2. Soil fauna (micro fauna) animal like *e.g.* Protozoa, Nematodes, earthworms, moles, ants, rodents.

Relative proportion / percentage of various soil microorganisms are: Bacteria-aerobic (70%), anaerobic (13 %), Actinomycetes (13%), Fungi /molds (03 %) and others (Algae Protozoa viruses) 0.2-0.8 %. Soil organisms play key role in the nutrient transformations.

Soil Humus

Humus is the organic residue in the soil resulting from decomposition of plant and animal residues in soil, or it is the highly complex organic residual matter in soil which is not readily degraded by microorganism, or it is the soft brown/dark coloured amorphous substance composed of residual organic matter along with dead microorganisms.

Composition of Humus

In most soil, percentage of humus ranges from 2-10 percent, whereas it is up to 90 percent in peat bog. On average humus is composed of Carbon (58 %), Nitrogen (3-6 %, Av.5%), acids - humic acid, fulvic acid, humin, apocrenic acid, and C: N ratio 10:1 to 12:1. During the course of their activities, the microorganisms synthesize number of compounds which plays important role in humus formation.

Functions/Role of Humus

1. It improves physical condition of soil
2. Improve water holding capacity of soil
3. Serve as store house for essential plant nutrients
4. Plays important role in determining fertility level of soil
5. It tend to make soils more granular with better aggregation of soil particles
6. Prevent leaching losses of water soluble plant nutrients

7. Improve microbial/biological activity in soil and encourage better development of plant-root system in soil
8. Act as buffering agent *i.e.* prevent sudden change in soil PH/soil reaction
9. Serve as source of energy and food for the development of soil organisms
10. It supplies both basic and acidic nutrients for the growth and development of higher plants
11. Improves aeration and drainage by making the soil more porous

Distinct Phases of Soil Microbiology

1. Ecological phase

Study on the quantitative and qualitative composition of the microscopic and ultramicroscopic soil population.

2. Experimental or physiological phase

Study on the physiology and biochemistry of the organisms, their role in the cycle of life in nature and their utilization for the formation of valuable metabolic products.

3. Agronomical phase

Application of microbiological activities for the enhancement of soil fertility and crop production.

4. Pedological phase

Importance of microorganisms in soil formation and soil structure.

Scope and Importance of Soil Microbiology

Living organisms both plant and animal types constitute an important component of soil. Though these organisms form only a fraction (less than one percent) of the total soil mass, but they play important role in supporting plant communities on the earth surface. While studying the scope and importance of soil microbiology, soil-plant-animal ecosystem as such must be taken into account. Therefore, the scope and importance of soil microbiology, can be understood in better way by studying aspects like

1. Soil as a living system
2. Soil microbes and plant growth
3. Soil microorganisms and soil structure
4. Organic matter decomposition
5. Humus formation
6. Biogeochemical cycling of elements
7. Soil microorganisms as bio-control agents
8. Soil microbes and seed germination

9. Biological N_2 fixation
10. Degradation of pesticides in soil.

1. **Soil as a living system:** Soil inhabit diverse group of living organisms, both micro flora (fungi, bacteria, algae and actinomycetes) and micro-fauna (protozoa, nematodes, earthworms, moles, ants). The density of living organisms in soil is very high *i.e.* as much as billions / gm of soil, usually density of organisms is less in cultivated soil than uncultivated / virgin land and population decreases with soil acidity. Top soil, the surface layer contains greater number of microorganisms because it is well supplied with Oxygen and nutrients. Lower layer / subsoil is depleted with Oxygen and nutrients hence it contains fewer organisms. Soil ecosystem comprises of organisms which are both, autotrophs (Algae, BOA) and heterotrophs (fungi, bacteria). Autotrophs use inorganic carbon from CO_2 and are "primary producers" of organic matter, whereas heterotrophs use organic carbon and are decomposers/consumers.
2. **Soil microbes and plant growth:** Microorganisms being minute and microscopic, they are universally present in soil, water and air. Besides supporting the growth of various biological systems, soil and soil microbes serve as a best medium for plant growth. Soil fauna & flora convert complex organic nutrients into simpler inorganic forms which are readily absorbed by the plant for growth. Further, they produce variety of substances like IAA, gibberellins, antibiotics *etc.* which directly or indirectly promote the plant growth
3. **Soil microbes and soil structure:** Soil structure is dependent on stable aggregates of soil particles-Soil organisms play important role in soil aggregation. Constituents of soil are *viz.* organic matter, polysaccharides, lignins and gums, synthesized by soil microbes plays important role in cementing / binding of soil particles. Further, cells and mycelial strands of fungi and actinomycetes, Vormicasts from earthworm is also found to play important role in soil aggregation.

   Different soil microorganisms, having soil aggregation / soil binding properties are graded in the order as Fungi > Actinomycetes > Gum producing Bacteria > Yeasts. Examples are: Fungi like Rhizopus, Mucor, Chaetomium, Fusarium, Cladasporium, Rhizoctonia, Aspergillus, Trichoderma and Bacteria like Azofobacler, Rhizobium Bacillus and Xanlhomonas.
4. **Soil microbes and organic matter decomposition:** The organic matter serves not only as a source of food for microorganisms but also supplies energy for the vital processes of metabolism that are characteristics of living beings. Microorganisms such as fungi, actinomycetes, bacteria, protozoa *etc.* and macro organisms such as earthworms, termites, insects *etc.* plays important role in the process of decomposition of organic matter and release of plant nutrients in soil. Thus, organic matter added to the soil is converted by oxidative decomposition to simpler nutrients / substances for plant growth and the residue is transformed into humus. Organic matter / substances include cellulose, lignins and proteins (in cell wall

The numbers of soil microorganisms carry out the process of biological nitrogen fixation at normal atmospheric pressure (1 atmosphere) and temperature (around 20 °C).

Two groups of microorganisms are involved in the process of Biological Nitrogen Fixation.

A. Non-symbiotic (free living)

B. Symbiotic (Associative)

A. Non-symbiotic (free living)

Depending upon the presence or absence of oxygen, non symbiotic N_2 fixation prokaryotic organisms may be aerobic heterotrophs (Azotobacter, Pseudomonas, Achromobacter) or aerobic autotrophs (Nostoc, Anabena, Calothrix, BGA) and anaerobic heterotrophs (Clostridium, Kelbsiella. Desulfovibrio) or anaerobic Autotrophs (Chlorobium, Chromnatium, Rhodospirillum, Meihanobacterium *etc*)

B. Symbiotic (Associative)

The organisms involved are Rhizobium, Bratfyrhizobium in legumes (aerobic): Azospirillum (grasses), Actinonycetes Frankia(with Casuarinas, Alder).

Soil microbes as biocontrol agents: Several ecofriendly bioformulations of microbial origin are used in agriculture for the effective management of plant diseases, insect pests, weeds *etc.* eg: Trichoderma sp and Gleocladium sp are used for biological control of seed and soil borne diseases. Fungal genera Entomophthora, Beauveria, Metarrhizium and protozoa Maltesia grandis. Malameba locustiae *etc* are used in the management of insect pests. Nuclear polyhydrosis virus (NPV) is used for the control of Heliothis / American boll worm. Bacteria like Bacillus thuringiensis, Pseudomonas are used in cotton against Angular leaf spot and boll worms.

Degradation of pesticides in soil by microorganisms: Soil receives different toxic chemicals in various forms and causes adverse effects on beneficial soil micro flora / micro fauna, plants, animals and human beings. Various microbes present in soil act as the scavengers of these harmful chemicals in soil. The pesticides/chemicals reaching the soil are acted upon by several physical, chemical and biological forces exerted by microbes in the soil and they are degraded into non-toxic substances and thereby minimize the damage caused by the pesticides to the ecosystem. For example, bacterial genera like Pseudomonas, Clostridium, Bacillus, Thiobacillus, Achromobacter *etc.* and fungal genera like Trichoderma, Penicillium, Aspergillus, Rhizopus, and Fusarium are playing important role in the degradation of the toxic chemicals / pesticides in soil.

Biodegradation of hydrocarbons: Natural hydrocarbons in soil like waxes, paraffin's, oils *etc* are degraded by fungi, bacteria and actinomycetes. *E.g.* ethane (C_2H_6) a paraffin hydrocarbon is metabolized and degraded by Mycobacteria, Nocardia, Streptomyces Pseudomonas, Flavobacterium and several fungi.

plant and from that plant infect another healthy plant, continuing this process indefinitely - therefore the infective agent had to be capable of reproduction. He believed that the filtered juice of the infected plants was itself alive, and he called the causative agent a filterable virus (the Latin word for poison). First isolated several soil bacteria *viz.,Azotobacter chroococcum, Thiobacillus thioparus.* Isolated first *Azospirillum* from *Paspalum* roots in Brazil and named it as *Azotobacter paspali* in 1922. He developed **enrichment culture** technique, a way to isolate microbes with certain growth preferences. Using enrichment media, it was possible to isolate enormous variety of bacteria with different physiological capabilities and different growth ranges. This demonstrated the impressive variety of microbial lifestyles, much broader than that of animals or plants.

Beijerinck and Winogradsky (1890) developed the enrichment culture technique for isolation of soil organisms, proved independently that transformation of nitrogen in nature is largely due to the activities of various groups of soil microorganisms (1891). Therefore, they are considered as «**Pioneer›s in soil bacteriology".**

Sergei Winogradsky (1856-1953)

Winogradsky extended awareness of microbial diversity. He discovered bacteria capable of **autotrophic** ("self-feeding") growth using inorganic compounds such as H_2S as their only energy source, and CO_2 as their only C-source. This was the first discovery that organisms other than green plants or algae could exist without consuming organic matter. Also discovered groups of photosynthetic bacteria that do not produce oxygen as their waste product. Developed culture technique known as **Winogradsky column**, in which mud and water are left in glass tube exposed to light. Over time, different microbial communities grow and interchange waste products and nutrients. He Established N & S transformation in soil.1890 and demonstrated the role of bacteria in nitrification.**S. N. Winogadsky (1891)** demonstrated the role of bacteria in nitrification and further in fill 1983 demonstrated that free living *Clostridium pasteurianum* could fix atmospheric nitrogen.

Omeliansky (1902) found the anaerobic degradation of cellulose by soil bacteria.

J. G. Lipman and P. E. Brown (1903, USA) studied ammonification of organic nitrogenous substances by soil microorganisms and developed the **Tumbler or Beaker** for studying different types of transformation in soil.

Hiltner (Germany, 1904) coined the term **"Rhizosphere"** to denote that region of soil which is subjected to the influence of plant roots. Rhizosphere is the region where soil and plant roots make contact.

Russel and Hutchinson (1909, England), proved the importance of protozoa controlling/ maintaining bacterial population and their activity in soil.

Conn (1918) developed **"Direct soil examination"** technique for studying soil microorganisms.

The numbers of soil microorganisms carry out the process of biological nitrogen fixation at normal atmospheric pressure (1 atmosphere) and temperature (around 20 °C).

Two groups of microorganisms are involved in the process of Biological Nitrogen Fixation.

A. Non-symbiotic (free living)

B. Symbiotic (Associative)

A. Non-symbiotic (free living)

Depending upon the presence or absence of oxygen, non symbiotic N_2 fixation prokaryotic organisms may be aerobic heterotrophs (Azotobacter, Pseudomonas, Achromobacter) or aerobic autotrophs (Nostoc, Anabena, Calothrix, BGA) and anaerobic heterotrophs (Clostridium, Kelbsiella. Desulfovibrio) or anaerobic Autotrophs (Chlorobium, Chromnatium, Rhodospirillum, Meihanobacterium *etc*)

B. Symbiotic (Associative)

The organisms involved are Rhizobium, Bratfyrhizobium in legumes (aerobic): Azospirillum (grasses), Actinonycetes Frankia(with Casuarinas, Alder).

Soil microbes as biocontrol agents: Several ecofriendly bioformulations of microbial origin are used in agriculture for the effective management of plant diseases, insect pests, weeds *etc*. eg: Trichoderma sp and Gleocladium sp are used for biological control of seed and soil borne diseases. Fungal genera Entomophthora, Beauveria, Metarrhizium and protozoa Maltesia grandis. Malameba locustiae *etc* are used in the management of insect pests. Nuclear polyhydrosis virus (NPV) is used for the control of Heliothis / American boll worm. Bacteria like Bacillus thuringiensis, Pseudomonas are used in cotton against Angular leaf spot and boll worms.

Degradation of pesticides in soil by microorganisms: Soil receives different toxic chemicals in various forms and causes adverse effects on beneficial soil micro flora / micro fauna, plants, animals and human beings. Various microbes present in soil act as the scavengers of these harmful chemicals in soil. The pesticides/chemicals reaching the soil are acted upon by several physical, chemical and biological forces exerted by microbes in the soil and they are degraded into non-toxic substances and thereby minimize the damage caused by the pesticides to the ecosystem. For example, bacterial genera like Pseudomonas, Clostridium, Bacillus, Thiobacillus, Achromobacter *etc*. and fungal genera like Trichoderma, Penicillium, Aspergillus, Rhizopus, and Fusarium are playing important role in the degradation of the toxic chemicals / pesticides in soil.

Biodegradation of hydrocarbons: Natural hydrocarbons in soil like waxes, paraffin's, oils *etc* are degraded by fungi, bacteria and actinomycetes. *E.g.* ethane (C_2H_6) a paraffin hydrocarbon is metabolized and degraded by Mycobacteria, Nocardia, Streptomyces Pseudomonas, Flavobacterium and several fungi.

2

History and Development of Soil Microbiology

History of Soil Microbiology (1600 - 1920)

There is enough evidence in the literature to believe that microorganisms were the earliest of the living things that existed on this planet. Man depends on crop plants for his existence and crop plants in turn depend on soil and soil microorganisms for their nutrition. Scientists form the beginning studied the microorganisms from water, air, soil *etc.* and recognized the role of microorganisms in natural processes and realized the importance of soil microorganisms in growth and development of plants.

Thus, we see that microorganisms have been playing a significant role long before they were discovered by man. Today, soil is considered to be the main source of scavenging the organic wastes through microbial action and is also a rich store house for industrial micro flora of great economic importance.

Unlike soil science whose origin can be traced back to Roman & Aryan times, soil microbiology is emerged as a distinct branch of soil science during first half of the 19th century. Some of the notable contributions made by several scientists in field of soil microbiology are highlighted in the following paragraphs.

A. V. Leeuwenhock (1673) discovered and described microorganisms through his own made first simple microscope with magnification of 200 to 300 times. He observed minute, moving objects which he called **"animalcules"** (small animals) which are now known as protozoa, fungi and bacteria. He for the first time made the authentic drawings of microorganisms (protozoa, bacteria, fungi).

Robert Hook (1635-1703) developed a compound microscope with multiple lenses and described the fascinating world of the microbes.

J. B. Boussingault (1838) showed that leguminous plants can fix atmospheric nitrogen and increase nitrogen content in the soil.

J. Von Liebig (1856) showed that nitrates were formed in soil due to addition of nitrogenous fertilizers in soil.

S. N. Winogradsky discovered the autotrophic mode of life among bacteria and established the microbiological transformation of nitrogen and sulphur. Isolated for the first time nitrifying bacteria and demonstrated role of these bacteria in nitrification (1890), further he demonstrated that free-living *Clostridium pasteuriamum* could fix atmospheric nitrogen (1893). Therefore, he is considered as **"Father of Soil Microbiology"**.

W. B. Leismaan (1858) and M. S. Woronin (1866) demonstrated that root nodules in legumes were formed by a specific group of bacteria.

Jodin (1862, France) gave the first experimental evidence of elemental nitrogen fixation by microorganisms.

Robert Koch (1882) developed gelatin plate/ streak plate technique for isolation of specific type of bacteria in soil, formulated Koch›s postulates to establish causal relationship between host - pathogen and disease.

R. Warington (1878) showed that nitrification in soil was a microbial process.

B. Frank i) discovered (1880) an actinomycetes "Frankia" (Actinorhizal symbiosis) inducing root nodules in non-legumes tress of genera *Alnus sp* and *Casurina* growing in temperate forests, ii) coined (1885) the term « Mycorrhiza» to denote association of certain fungal symbionts with plant roots (Mycorrhiza-A symbiotic association between a fungus and roots of higher plants. Renamed the genus Bacillus as Rhizobium (1889).

H. Hellriegel and H. Wilfarth (1886) showed that the growth of non-legume plant was directly proportional to the amount of nitrogen supplied, whereas, in legumes there was no relationship between the quantity of nitrogen supplied and extent of plant growth. They also suggested that bacteria in the root nodules of legumes accumulate atmospheric nitrogen and made it available to plants. Showed that a mutually beneficial association exists between bacteria (*Rhizobia*) and legume root and legumes could utilize atmospheric nitrogen (1988).

Beijerinck, Martinus Willem (1851-1931)

M. W. Beijerinck (1888) isolated root nodule bacteria in pure culture from nodules in legumes and named them as *Bacillus radicola* Considered as **father of "Microbial ecology"**. He was the first Director of the Delft School of microbiology (Netherland).

Beijerinck was born in Amsterdam and studied chemical engineering at the Delft Polytechnic, where, after working as a bacteriologist with an industrial company, he taught and carried out research from 1895.He published his finding in 1898 that an agent smaller than bacteria could cause diseases, an agent that he called a virus. In the 1880s and 1890s Beijerinck studied the disease that stunts the growth of tobacco plants and mottles their leaves (now called the tobacco mosaic virus disease). Trying to isolate a causative agent, he pressed out the juice of infected tobacco leaves and found that the juice alone was able to infect healthy plants, even after he had passed it through a filter that removed bacteria. Beijerinck was also certain that the causative agent was not a toxin because he could infect a healthy

plant and from that plant infect another healthy plant, continuing this process indefinitely - therefore the infective agent had to be capable of reproduction. He believed that the filtered juice of the infected plants was itself alive, and he called the causative agent a filterable virus (the Latin word for poison). First isolated several soil bacteria *viz.,Azotobacter chroococcum, Thiobacillus thioparus.* Isolated first *Azospirillum* from *Paspalum* roots in Brazil and named it as *Azotobacter paspali* in 1922. He developed **enrichment culture** technique, a way to isolate microbes with certain growth preferences. Using enrichment media, it was possible to isolate enormous variety of bacteria with different physiological capabilities and different growth ranges. This demonstrated the impressive variety of microbial lifestyles, much broader than that of animals or plants.

Beijerinck and Winogradsky (1890) developed the enrichment culture technique for isolation of soil organisms, proved independently that transformation of nitrogen in nature is largely due to the activities of various groups of soil microorganisms (1891). Therefore, they are considered as **«Pioneer»s in soil bacteriology".**

Sergei Winogradsky (1856-1953)

Winogradsky extended awareness of microbial diversity. He discovered bacteria capable of **autotrophic** ("self-feeding") growth using inorganic compounds such as H_2S as their only energy source, and CO_2 as their only C-source. This was the first discovery that organisms other than green plants or algae could exist without consuming organic matter. Also discovered groups of photosynthetic bacteria that do not produce oxygen as their waste product. Developed culture technique known as **Winogradsky column**, in which mud and water are left in glass tube exposed to light. Over time, different microbial communities grow and interchange waste products and nutrients. He Established N & S transformation in soil.1890 and demonstrated the role of bacteria in nitrification.**S. N. Winogadsky (1891)** demonstrated the role of bacteria in nitrification and further in fill 1983 demonstrated that free living *Clostridium pasteurianum* could fix atmospheric nitrogen.

Omeliansky (1902) found the anaerobic degradation of cellulose by soil bacteria.

J. G. Lipman and P. E. Brown (1903, USA) studied ammonification of organic nitrogenous substances by soil microorganisms and developed the **Tumbler or Beaker** for studying different types of transformation in soil.

Hiltner (Germany, 1904) coined the term **"Rhizosphere"** to denote that region of soil which is subjected to the influence of plant roots. Rhizosphere is the region where soil and plant roots make contact.

Russel and Hutchinson (1909, England), proved the importance of protozoa controlling/ maintaining bacterial population and their activity in soil.

Conn (1918) developed **"Direct soil examination"** technique for studying soil microorganisms.

Rayner (192I) and Melin (1927) carried out the intensive study on Mycorrhiza.

The fertility of soil depends not only on its chemical composition but also on the qualitative and quantitative nature of microorganisms inhabiting it. Strictly speaking, the development of microbiology as a branch of science can be dated back to the time of people who ground lens from glass and saw microorganisms, through them.

Anton Van Leeuwenhoek, a linen draper from Holland (1632-1723) is credited with having made the first authenticated drawings of microorganisms which had its roots in the age of Aristotle (384-322 B.C) and which was revoked by the so-called experimental evidence of John Needham (1713-1781) was scattered by the conclusive experimental findings of Louis Pasteur (1822-1895). This was followed by Robert Koch's Koch postulates (1843-1910) and the pure culture technique of Joseph Lister (1878).

Soil microbiology emerged a distinct branch of soil science only in 1838 after the French agricultural chemist and farmer, J.B. Boussingault showed that legumes can obtain nitrogen from air when grown in soil which was not heated. Since then several significant developments made in soil microbiology are in the following.

Selman.A.Waksman (1944)

In 1922, Waksman isolated *Thiobacillus thioxidans.* He discovered the antibiotic Streptomycin in the year 1944 for tuberculosis - *Mycobacterium tuberculosis.* Also discovered Neomycin, Actinomycin antibiotics. He wrote a book on Principles of Soil Microbiology. In 1942, he showed the importance of soil as the source of antagonistic organisms. He also studied variety of biochemical reactions carried out by soil microorganisms while decomposing organic matter.

S. A. Waksman published the book **"Principles of soil Microbiology"** and thereby encouraged the research in soil microbiology (1927). Studied the role of soil as the source of antagonistic organisms with special reference to soil actinomycetes (1942) and discovered the antibiotic «Streptomycin» produced by *Streptomyces griseus*, a soil actinomycets (1944).

Rossi (1929) and Cholondy (1930) developed **"Contact Slide / Buried slide"** technique for studying soil micro flora.

Van Niel (1931USA) studied chemoautotrophic bacteria and bacterial photosynthesis.

Bortels (1936) demonstrated the importance of molybdenum in accelerating nitrogen fixation by nodulating legumes.

Garrett (1936) established the school in UK on «Soil fungi and ecological classification».

Kubo (1939, Japan) showed / proved-the role and importance of "leghaemoglobin" (Red pigment) present in root nodules of legumes in nitrogen fixation.

Ruinen (1956) Dutch microbiologist coined the term **"Phyllosphere"** to denote the region of leaf influenced by microorganisms.

Alien et al (1980) (suggested that VAM fungi stimulate plant growth by physiological effects other than by enhancement of nutrient uptake.

Jensen (1942) developed the method of studying nodulation on agar media in test tubes.

Barbara Mosse and J. W. Gerdemann (1944) reported occurrence of VAM (vesicular-arbuscular Mycorrhiza) fungi (*Glomus, Aculopora genera*) in the roots of agricultural crop plants which helps in the mobilization of phosphate.

Starkey (1945) studied role of bacteria (Bacillus and Clostridium) in the transformation of iron.

Barker (1945) studied anaerobic fermentation by methane bacteria (*Methanococcus, Methanosarcina*)

Thornton, (1947), studied root nodule bacteria form clovers.

Virtanen (1947) studied chemistry and mechanism of leghaemoglobin in nitrogen fixation.

Nutman (1948 England) studied hereditary mechanism of root nodulation in legumes.

Burris and Wilson (1957) developed the **"Isotope technique"** to quantify the amount of nitrogen fixed and further isolated and characterized the enzyme "Nitrogenase".\

Bergersen (1957 Australia) elaborated the biochemistry of nitrogen fixation in legume root nodules.

Carnham (1960 USA) discovered nitrogen fixation by cell-free extract of *Clostridium pasteurianum.*

Alexander Fleming (1881 – 1955)

In 1928, while working on influenza virus, he observed that mould had developed accidentally on a staphylococcus culture plate and that the mould had created a bacteria-free circle around itself. He was inspired to further experiment and he found that a mould culture prevented growth of staphylococci, even when diluted 800 times. In 1929 he discovered the Antibiotic *Penicillin,* which is the important milestone in medical microbiology. Florey and Chain later isolated penicillin in pure culture. Alexander Fleming got the Nobel Prize in medicine in the year 1945. He found that natural substances / natural products are having antimicrobial activity. He reported that Nasal mucous, saliva are having antimicrobial property due to the action of lysozyme. He started the **"School of soil microbiology"** at Cornell University to study microbial aspects of pesticides degradation (1961) and developed the antibiotic «Penicillin» from the fungus *Penicillium notatum* (1929).

soil. From the beginning of 20th century emphasis was given to the study of microorganisms in soil in relation to their physiology, ecology, interrelationship, role in soil processes and soil fertility. Further role of fungi and actinomycetes in cellulose decomposition was better understood and cellulose decomposing, sulphur oxidizing, iron bacteria *etc* were isolated from soil and studied in detail.

History of Soil Microbiology in India

During last few decades greater emphasis has been given on some of the important aspects in soil microbiology in India which are:

1. Characterization of N_2 fixing Azotobacter, Rhizobium, Beijerinckia, BGA *etc.*
2. Studies on P- solubilizing bacteria and fungi, celluloytic microorganisms, silage production role of humic acid *etc.*
3. Establishment (1979) of All-India Coordinated Project (AICP) on BNF at IARI and field oriented work on BNF.
4. Standardization of methods of bio-inoculants application to seed and soil.
5. Seed bacterization and response of crops to bio-inoculants.

Some of the most important contributions made on the different aspects in the field of soil microbiology by the scientists and research institutes in the country are highlighted in the following paragraphs:

C. N. Acharya (1940) contributed towards the better utilization of Agricultural wastes for the production of biogas & compost.

Sundara Rao (1962) established the **"Division of Microbiology"** at IARI New Delhi.

Madhok (Punjab) introduced the practice of using bacterial cultures for berseem.

Sanyasi Raju & Rajagopalan (Coimbatore) initiated the research work on root nodulation in legumes at Madras, Agil. College.

P. K. Dey (West Bengal) worked on free living N2 fixing organisms *viz Azotobacter, Beijerinckia* and BGA in rice fields and discovered N2 fixation by BGA in paddy.

M.O.P. lyengar (Madras Univ.) laid foundation stone of algal research in India.

Sadasivan (Madras) and Saxena (Allahabad) studied ecology and physiology of soil fungi along with rhizosphere phenomenon.

Singh B.N. pioneering research on soil protozoa in India.

Bhar J.V. (Bangalore) initiated work on the role of earthworms in the maintenance of soil fertility, biological nitrogen fixation and microbiology of phyllosphere.

Thirumalacher (Hindustan Antibiotics, Pune) developed antifungal antibiotics like Haymycin and Aureofungin.

Nandi (Bose Res. Institute, Calcutta) worked on production technology of antibiotics and bacterial fertilizers (Biofertilizers).

Desikachray (Madras) studied taxonomy of BGA in India.

Thomas (BARC, Mumbai) studied physiology of algae in India.

Raja Rammohan Rao (CRRI, Cuttak) studied on rhizosphere nitrogen fixation phenomenon.

Bhagyaraj (GKVK, Bangalore) studied Mycorrhiza and N2 fixation interactions.

Verma (JNU, Delhi) studied / worked on sulphur metabolism.

Subramaniam & Mahadevan (Univ. Madras) studied fundamental aspects of N2 fixation.

Modi, Sushil Kumar, Das and Thomas carried research on "Genetics of "Nif" gene in relation to BNF by *Rhizobium, Azospirillum* and *Kelbsiella.*

Bharadwaj (Palampur) studied / worked on microbiology of organic matter decomposition & role of celluloytic microorganisms.

Gaur (IARI) and Mishra (Hissar) studied the role of celluloytic microorganisms in accelerating the process of composting and compost making.

Karla and Garcha (Ludhiana) studied the phenomenon of cellulose degradation and legume bacteriology.

Ranganathan & Nellakantan (NDRI, Karnal) worked on silage microbiology and process of anaerobic decomposition in biogas production.

Vadher, Gupta, Sethunatathan and Raghu studied role of soil enzymes and microbiology of pesticide degradation in soil.

Dart & Wani (non symbiotic N_2 fixation), **Thomas, Kumar Rao, Nambiar and Rupela** (symbiotic N_2 fixation) and **Krishna** (VAM fungi), these scientist at ICRISAT, Hyderabad work on symbiotic & non-symbiotic N_2 fixation in Gram, Groundnut, Arhar, Sorghum and Millets.

N. V. Joshi (1920) reported first isolation and identification *of Rhizobium* from different cultivated legumes

Gangulee and Madhok, studied physiology of *Rhizobium* and production of *Rhizobium* inoculants.

Sen and Pal (1957) studied solubilization of phosphate by soil microorganisms.

A. Sankaran (1958) standardized quality of legume inoculants for first time in India.

P. K. Dey and R. Bhattacharya isolated for the first time a new, non-symbiotic N2 fixing bacterium *Derixa gummosa* in the world.

V. Iswaran (1959) reported the use of Indian peat as carrier for Biofertilizers production.

Dube J. N. (1975) reported coal (wood-coal), an alternative to peat as carrier material for biofertilizer production.

3

Distribution and Importance of Microorganisms

Soil is a bioorganic - mineral complex of the loose surface layer of the earth's crust. Soil microbiology is the science that deals with the millions of microscopic forms of life found in the soil. It is stated that one hectare - foot (one hectare to a depth of 30cm) contains 2 to 10 metric tons of living microorganisms. Soil microorganisms are in a constant struggle for existence and are believed to be in a state of dynamic equilibrium. Without microorganisms, soil would be an inactive geological mass incapable of supporting plant life and without plant life, animal life would not be possible. This soil population consists mainly of bacteria, fungi, actinomycetes, algae and protozoa. The soil ecosystem includes these microbial groups as well as the inorganic and organic constituents of a given site. The collections of cells represented in the community are considered as distinct populations. All the inhabitants of the particular locality make up the community.

Types of Microorganisms in Soil

Living organisms both plants and animals, constitute an important component of soil. The pioneering investigations of a number of early microbiologists showed for the first time that the soil was not an inert static material but a medium pulsating with life. The soil is now believed to be a dynamic or rather a living system, containing a dynamic population of organisms/microorganisms. Cultivated soil has relatively more population of microorganisms than the fallow land, and the soils rich in organic matter contain much more population than sandy and eroded soils. Microbes in the soil are important to us in maintaining soil fertility / productivity, cycling of nutrient elements in the biosphere and sources of industrial products such as enzymes, antibiotics, vitamins, hormones, organic acids *etc*. At the same time certain soil microbes are the causal agents of human and plant diseases. The soil organisms are broadly classified in to two groups *viz* soil flora and soil fauna, the detailed classification of which is as follows.

Soil Organisms

A. Soil Flora

a) Microflora: 1. Bacteria 2. Fungi, Molds, Yeast, Mushroom 3. Actinomycetes 4.Algae.

b) Macroflora: Roots of higher plants

B. Soil Fauna

a) Microfauna: Protozoa, Nematodes

b) Macrofauna: Earthworms. moles, ants & others.

As soil inhabit several diverse groups of microorganisms, but the most important amongst them are: bacteria, actinomycetes, fungi, algae and protozoa. The characteristics and their functions / role in the soil are described in the next topics.

I. Bacteria

Bacteria are the most dominant group of M.O in soil and more numerous than the other four combined. They present in all types of soil but their population decreases as the depth of soil increases (Horizon A > B > C). The number of cells of bacteria in the soil is always great, but the individuals are small, (µm in length). Because of the minute size of the bacteria and large cells or filaments of the other 4 groups, the bacteria probably account for appreciably less than half of the total microbial cell mass. In aerated soil, B, will be dominating, alone are responsible for all the activities in environment lacking O_2 or little O_2. In transformation process Bacteria stand first, due to their rapid growth and capacity of vigorous decomposition of variety of substrates.

Soil microbiological population has been divided into two broad groups. 1. Autohcthonous 2. Zymogenous.

Autochthonous or native microbes; Indigenous, which are characteristic of the particular soil and always found to be there.*e.g.* Arthrobacter

Zymogenous, or fermentative organisms require an external source of and their normal population in soil is low (Pseudomonas, Bacillus). When specific substrates are added to soil, the population is increased. Then gradually declines when the added substrate is exhausted. *e.g.* Cellulose decomposers. N utilizing bacteria, nitrifiers *etc.*,

All the bacteria essentially need carbon, energy and electron for their normal cellular activity, growth and reproduction. Based on the source of above, bacteria can be classified in to following groups.

1. **Autotrophs:** They utilize CO_2 as sole source for carbon requirement. Autotrophic bacteria are capable synthesizing their food from simple inorganic nutrients, while heterotrophic bacteria depend on pre-formed food for nutrition. All autotrophic bacteria utilize CO_2 (from atmosphere) as carbon source and derive energy either from sunlight (photoautotrophs, *e.g.* Chromatrum. Chlorobium. Rhadopseudomonas or from the oxidation of simple inorganic substances present in soil (chemoautotrophs *e.g.* Nitrobacter, Nitrosomonas, Thiaobacillus).
2. **Heterotrophs:** They derive their carbon from organic compounds.Majority of soil bacteria are heterotrophic in nature and derive their carbon and energy from complex organic substances/organic matter, decaying roots and plant residues. They obtain their nitrogen from nitrates and ammonia

compounds (proteins) present in soil and other nutrients from soil or from the decomposing organic matter. Certain bacteria also require amino acids, B- Vitamins, and other growth promoting substances also.*e.g.* Symbiotic & Non - Symbiotic N_2 fixers, Ammonifier, Cellulose Decomposers, Denitrifiers

3. **Phototrophs:** Bacteria capable of deriving energy from sunlight.
4. **Chemotrophs:** They utilize inorganic chemicals for their energy.
5. **Lithotroph:** The organisms utilize inorganic substances as electron source.
6. **Organotroph:** Bacteria capable of utilizing organic substances as electron source.

They may exist in combinations of the above groups as follows. Photolithotrophs; Chemolithotrophs; Photoorganotrops; Chemoorganotrophs; Photoautotrophs; Photoheterotrophs; Chemoautotrophs; Chemoheterotrophs.

Abundance of Bacteria

In Arctic regions where in the frozen condition 9-10 months / year bacterial counts of 10^6 / g are observed even when temperature remains below freezing point. Such organisms are in a state of dormancy awaiting the spring to recover. Desert soil is another extreme. Bacteria exist even in oven dry state. Spore forming bacilli are often predominant in this environment. Common soil bacteria are Arthrobacter, Bacillus, Clostridium, Micrococcus.

As per the system proposed in the Bergey's Manual of Systematic Bacteriology, most of the bacteria which are predominantly encountered in soil are taxonomically included in the three orders, Pseudomonadales, Eubacteriales and Actinomycetales of the class Schizomycetes. The most common soil bacteria belong to the genera Pseudomonas, Arthrobacter, Clostridium Achromobacter, Sarcina, Enterobacter *etc*. The another group of bacteria common in soils is the Myxobacteria belonging to the genera Micrococcus, Chondrococcus, Archangium, Polyangium, Cyptophaga.

Importance

1. In transformation processes, bacteria stand first due to their rapid growth and capacity of vigorous decomposition of variety of substrates.
2. Most of the organisms are involved in N_2 fixation, P solubilization, CO_2 fixation, S and Fe transformation, Si solubilization *etc*.,
3. They also act as molecular vectors.

Functions / Role of Bacteria

Bacteria bring about a number of changes and biochemical transformations in the soil and thereby directly or indirectly help in the nutrition of higher plants growing in the soil. The important transformations and processes in which soil bacteria play vital role are: decomposition of cellulose and other carbohydrates, ammonification (proteins ammonia), nitrification (ammonia-nitrites-nitrates), denitrification (release of free elemental nitrogen), biological fixation of atmospheric nitrogen (symbiotic and non-symbiotic) oxidation and reduction of sulphur and iron compounds. All these processes play a significant role in plant nutrition,

Process/reaction	Bacterial genera
Cellulose decomposition (Celluloytic bacteria) most cellulose decomposers are mesophilic	a. Aerobic : *Angiococcus, Cytophaga, Polyangium, Sporocytophyga, Bacillus, Achromobacter, Cellulomonas* *b. Anaerobic: Clostridium Methanosarcina, Methanococcus*
Ammonification (Ammonifiers)	*Bacillus, Pseudomonas*
Nitrification (Nitrifying bacteria)	*Nitrosomonas, Nilrobacter Nitrosococcus*
Denitrification (Denitrifies)	*Achromobacter, Pseudomonas, Bacillus, Micrococcus*
Nitrogen fixing bacteria	a. Symbiotic- Rhizobium, *Bradyrrhizobium* b. Non-symbiotic: aerobic–Azotobacter *Beijerinckia* (acidic soils), anaerobic-*Clostridium*

Bacteria capable of degrading various plant residues in soil:

Cellulose	Hemicelluloses	Lignin	Pectin	Proteins
Pseudomonas	Bacillus	Pseudomonas	Erwinia	Clostridium
Cytophaya	Vibrio	Micrococcus		Proteus
Spirillum	Pseudomonas	Flavobacteriumm		Pseudomonas
Actinomycetes	Erwinia	Xanthomonas		Bacillus
Cellulomonas		Streptomyces		

II. Fungi

As the important constituent group of the soil population, they are widely distributed in most well - cultivated soils. Fungi account for a large part of the total microbial population. Though they are not the major inhabitants, they do in fact makeup the significant part of he biomass because of the large diameter and extensive network of the filaments. Fungi exist in soil in the form of vegetative mycelium and spores. Characteristically the fungi possess a filamentous mycelium network of individual hyphal stands. The hypha itself is rather broad and has a diameter appreciably greater than that found in the common actinomycetes. In nature consider or asexual spores are abundant and widespread, the sexual spores relatively uncommon. In contrast with bacteria the high can be effectively differentiated into Genera and species on the basis of morphology.

Fungi in soil are present as mycelial bits, rhizomorph or as different spores. Their numb... s from a few thousand to a few -million per gram of soil. Soil fung... ntous mycelium composed of individual hyphae. The fungal ... tate /coenocytic (Mastigomycotina and Zygomycotina) or ... Basidiomycotina & Deuteromycotina).

...lance

... aerobic and heterotrophic, they require abundant supply ... tter in soil. Fungi are dominant in acid soils, because ... nducive / suitable for the existence of either bacteria or actinomycetes. The optimum pH range for fungi lies-between 4.5 to 6.5. They are also present in neutral and alkaline soils and some can even tolerate PH beyond 9.0

Several techniques have been developed for the study of the fungal flora. Plate count is burial slide techniques are used for estimation. Slight modification of pH and addition of Novobiocin and streptomycin inhibit bacteria and actinomycets.

Estimates of microbial density revealed that a few as 20,000 to as many as 1,000,000 fungal propagules per gram, the propagule being considered as any spore, or hyphal filament that is capable of giving rise to a colony. The length of the fungal mycelium has been reported to range from 10 to 100m per g surface soil, but varies up to 500 and sometimes in excess of 1000mt has also been obtained. Assuming the filament has an average diameter of 5μm and a specific gravity of 1.2 and taking the range of 10 to 100 m per gram. It would appear that the weight of fungi ranges from 500 to 5000 kg per ha of surface soil. Thus the filaments make up a significant part of the soil mass. They are the inhabitants of peats, flooded soils planted to rice, regions with low and extremely high salt contents, locations in many deserts as well as sites in Antarctica. More numerous in surface layers of well-aerated and dominant in acid soils. Common genera in soil are Aspergillus, Mucor, Penicillium Trichoderma, Alternaria, Rhizopus.

As observed by C.K. Jackson (1975), most commonly encountered genera of fungi in soil are; *Alternaria, Aspergillus, Cladosporium, Cephalosporium Botrytis, Chaetomium, Fusarium, Mucor, Penicillium, Verticillium, Trichoderma, Rhizopus, Gliocladium, Monilia, Pythium,* etc. Most of these fungal genera belong to the subdivision Deuteromycotina / Fungi imperfeacta which lacks sexual mode of reproduction.

Importance

- ❑ They degrade complex molecules like cellulose, hemicellulose, pectin, starch, lignin in the soil.
- ❑ Fungi improve the soil structure by forming soil aggregates.
- ❑ They cause pathogenecity in plants, animals and human beings
- ❑ Act as predator against soil protozoa. The hyphal penetration reduces mobility and total cessation of movement.
- ❑ Nematode trapping fungi like Arthrobotrys, Dacylaria, Dactylella, Harposporiumtrap and kill the nematodes.
- ❑ They also have beneficial association with plant roots *e.g.* Mycorrhizal fungi

Functions / Role of Fungi

1. Fungi plays significant role in soils and plant nutrition.
2. They plays important role in the degradation / decomposition of cellulose hemi cellulose, starch, pectin, lignin in the organic matter added to soil.
3. Lignin which is resistant to decomposition by bacte decomposed by fungi.

4. They also serve as food for bacteria.
5. Certain fungi belonging to sub-division Zygomycotina and Deuteromycotina are predaceous in nature and attack on protozoa & nematodes in soil and thus, maintain biological equilibrium in soil.
6. They also plays important role in soil aggregation and in the formation of humus.
7. Some soil fungi are parasitic and causes number of plant diseases such as wilts, root rots, damping-off and seedling blights *e.g. Pythium, Phyiophlhora, Fusarium, Verticillium* etc.
8. Number of soil fungi forms mycorrhizal association with the roots of higher plants (symbiotic association of a fungus with the roots of a higher plant) and helps in mobilization of soil phosphorus and nitrogen *e.g.* Endomycorrhiza - Glomus, *Gigaspora, Aculospora,* and Ectomycorrhiza - *Amanita, Boletus, Entoloma, Lactarius.*

III. Actinomycetes

They are unicellular like bacteria, but produce a mycelium which is non-septate (coenocytic) and more slender, tike true bacteria they do not have distinct cell-wall and their cell wall is without chitin and cellulose (commonly found in the cell wall of fungi). On culture media unlike slimy distinct colonies of true bacteria which grow quickly, actinomycetes colonies grow slowly, show powdery consistency and stick firmly to agar surface. They produce hyphae and conidia / sporangia like fungi. Certain actinomycetes whose hyphae undergo segmentation resemble bacteria, both morphologically and physiologically.

A transitional group between the simple Bacteria and Fungi are the actinomycetes. These are microorganisms producing slender, branded filaments that develop into a mycelium in all soil. They are classified under bacteria, but resemble fungi in some aspects.

- Produces mycelium with extensive branching
- Like Fungi many actinomycetes form aerial mycelium and conidia
- Growth in liquid medium does not resemble (which bacteria is turbid), but form clumps and pellets like Fungi.

Certain actinomycetes resemble Mycobacterium and Corynebacterium in all respects both morphologically and physiologically including susceptibility to virus attack.

Actinomycetes differ from fungi that they do not have chitin and cellulose which are commonly found in the cell walls of fungi.The colony characters are not similar to bacteria.Some species have flagella that resemble those of true Bacteria.Similarities in cell wall composition and sensitivity to antibacterial compounds. Actinomycetes are sensitive to antibacterial compounds and not antifungal compounds

Common genera in soils are Streptomyces (70%), Nocardia, Micromonaspora, Frankia. They elaborate musty odour, an odour reminiscent of freshly turned soil.

2. Organic residues / substances added soil are first attacked by bacteria and fungi and later by actinomycetes, because they are slow in activity and growth than bacteria and fungi.
3. They decompose / degrade the more resistant and indecomposable organic substance/matter and produce a number of dark black to brown pigments which contribute to the dark colour of soil humus.
4. They are also responsible for subsequent further decomposition of humus (resistant material) in soil.
5. They are responsible for earthy / musty odor / smell of freshly ploughed soils.
6. Many genera species and strains (*e.g.* Streptomyces if actinomycetes produce/synthesize number of antibiotics like Streptomycin, Terramycin, Aureomycin *etc.*
7. One of the species of actinomycetes *Streptomyces scabies* causes disease "Potato scab" in potato.

IV. Yeast

Yeasts are non-filamentous fungi. Their presence may be demonstrated in most of the soils. Common soil yeasts are Candida, Debaryomyces, Rhodotorula, Torulopsis. Yeasts have been found in comparable numbers in soils of Antarctica, grasslands, cultivated fields, and forests.

Function and Activity

- Involved in the degradation of complex molecules.
- They can utilize and degrade the major plant constituents – cellulose, hemicellulose, pectin, starch and lignin.
- Participate in humus formation from fresh organic residues.
- Carry out inorganic transformations and influences the formation of stable aggregates by means of hyphal penetration and the mechanical binding of the particles.

V. Algae

Algae are present in most of the soils where moisture and sunlight areavailable. Their number in soil usually ranges from 100 to 10,000 per gram of soil. They are photoautotrophic, aerobic organisms and obtain CO_2 from atmosphere and energy from sunlight and synthesize their own food. They are unicellular, filamentous or colonial. Soil algae are divided in to four main classes or phyla as follows:

1. Cyanophyta (Blue-green algae)
2. Chlorophyta (Grass-green algae)
3. Xanthophyta (Yellow-green algae)
4. Bacillariophyta (diatoms or golden-brown algae)

Out of these four classes / phyla, blue-green algae and grass-green algae are more abundant in soil. The green-grass algae and diatoms are dominant in the soils of temperate region while blue-green algae predominate in tropical soils. Green-algae prefer acid soils while blue green algae are commonly found in neutral and alkaline soils. The most common genera of green algae found in soil are: Chlorella, Chlamydomonas, Chlorococcum, Protosiphon *etc.* and that of diatoms are Navicula, Pinnularia. Synedra, Frangilaria.

Blue green algae are unicellular, photoautotrophic prokaryotes containing Phycocyanin pigment in addition to chlorophyll. They do not posses flagella and do not reproduce sexually. They are common in neutral to alkaline soils. The dominant genera of BGA in soil are: Chrococcus, Phormidium, Anabaena, Aphanocapra, Oscillatoria *etc.* Some BGA posses specialized cells know as **"Heterocyst"** which is the sites of nitrogen fixation. BGA fixes nitrogen (non-symbiotically) in puddle paddy/water logged paddy fields (20-30 kg/ha/season). There are certain BGA which possess the character of symbiotic nitrogen fixation in association with other organisms like fungi, mosses, liverworts and aquatic ferns Azolla, eg Anabaena-Azolla association fix nitrogen symbiotically in rice fields.

They are photosynthetic organisms and abundant in habitats, where adequate moisture and accessible light are available. Isolation can be done in liquid media and by plating with agar media. Because of their similarities to bacteria, the blue green algae (cyanobacteria) are classified together with bacteria.

Photoautotrophic nutrition

In the absence of light heterotrophy occur in some species of Chlorophyta, Cyanrphyta and Diatoms. These species use the oxidation of organic carbon to replace the light in supplying energy for anabolic process. (Facultative photoautotrophs)

Occurrence of this photosynthetic microorganism has been recorded in soils throughout the world. Algae are moderately adaptable to environmental changes; persist in unfavourable circumstances such as in alkaline and desert soils. Flooded paddy field provides conducive environment for the growth of algae. The common inhabitants are Anabaena, Nostoc, Calothrix and Oscillatoria.

Importance

- Involved in CO_2 fixation
- Colonizes the barren surface and corrode and weather rocks which contribute to soil formation
- Algal layer covers the rocks, on death it favors secondary colonizers
- Surface algal bloom reduces erosion losses probably by binding together with soil particles
- Improve soil structure, texture and add fertility to soil after decay
- Photosynthesis liberates 02 , provides 02 to the roots and other microbes
- N2 fixation and ammonia excretion
- N2 is released after decomposing of algal biomass

- Production of metabolites
- Single cell protein and β carotene production

Functions / Role of Algae or BGA

1. Plays important role in the maintenance of soil fertility especially in tropical soils.
2. Add organic matter to soil when die and thus increase the amount of organic carbon in soil.
3. Most of soil algae (especially BGA) act as cementing agent in binding soil particles and thereby reduce/prevent soil erosion.
4. Mucilage secreted by the BGA is hygroscopic in nature and thus helps in increasing water retention capacity of soil for longer time/period.
5. Soil algae through the process of photosynthesis liberate large quantity of oxygen in the soil environment and thus facilitate the aeration in submerged soils or oxygenate the soil environment.
6. They help in checking the loss of nitrates through leaching and drainage especially in un-cropped soils.
7. They help in weathering of rocks and building up of soil structure.

VI. Protozoa

The subterranean earth contains protozoa, earthworms, nematodes, and insects. Most abundant one is protozoa, the simple form of animal life. Most of the soil forms are flagellates, amoebae or ciliates. Primitive, unicellular and majority of them are dependent on preformed organic matter for their nutrition from soluble organic and inorganic substances, or by phagotrophic nutrition (direct feeding upon microbial cells). They are regulating the biological equilibrium in soil by engulfing the bacteria. They also participate in the decomposition of plant remains.

These are unicellular, eukaryotic, colourless, and animal like organisms (Animal kingdom). They are larger than bacteria and size varying from few microns to a few centimeters. Their population in arable soil ranges from 10,000 to 1,00,000 per gram of soil and are abundant in surface soil. They can withstand adverse soil conditions as they are characterized by **"cyst stage"** in their life cycle. Except few genera which reproduce sexually by fusion of cells, rest of them reproduces asexually by fission / binary fission. Most of the soil protozoa are motile by flagella or cilia or pseudopodia as locomotors organs.

Depending upon the type of appendages provided for locomotion, protozoa are

1. Rhizopoda (Sarcondia)
2. Mastigophora
3. Ciliophora (Ciliata)
4. Sporophora (not common Inhabitants of soil)

Class-Rhizopoda: Consists of protozoa without appendages usually have naked protoplasm without cell-wall, pseudopodia as temporary locomotory organs are present some times. Important genera are Amoeba, Biomyxa, Euglypha, *etc.*

Class Mastigophora: Belongs to flagellated protozoa, which are predominant in soil. Important genera are: Allention, Bodo, Cercobodo, Cercomonas, Entosiphon Spiromonas, Spongomions and Testramitus. Many members are saprophytic and some posses chlorophyll and are autotrophic in nature. In this respect, they resemble unicellular algae and hence are known as "Phytoflagellates".

The soil protozoa belonging to the class ciliate / ciliophora are characterized by the presence of cilia (short hair-like appendages) around their body, which helps in locomotion. The important soil inhabitants of this class are Colpidium, Colpoda, Balantiophorus, Gastrostyla, Halteria, Uroleptus, Vortiicella, Pleurotricha *etc.*

Protozoa are abundant in the upper layer (15 cm) of soil. Organic manures protozoa. Soil moisture, aeration, temperature and PH are the important factors affecting soil protozoa.

Function / Role of Protozoa

1. Most of protozoans derive their nutrition by feeding or ingesting soil bacteria belonging to the genera *Enterobacter, Agrobacterium, Bacillus, Escherichia, Micrococcus,* and *Pseudomonas* and thus, they play important role in maintaining microbial / bacterial equilibrium in the soil.
2. Some protozoa have been recently used as biological control agents against phytopathogens.
3. Species of the bacterial genera *viz. Enterobacter* and *Aerobacter* are commonly used as the food base for isolation and enumeration of soil protozoans.
4. Several soil protozoa cause diseases in human beings which are carried through water and other vectors, *e.g.* Amoebic dysentery caused by *Entomobea histolytica*

Importance of soil microorganisms

Soil microorganisms are very important as almost every chemical transformation taking place in soil involves active contributions from these organisms. In particular, they play an active role in soil fertility as a result of their involvement in the cycle of nutrients like carbon and nitrogen, which are required for plant growth. For example, soil microorganisms are responsible for the decomposition of the organic matter entering the soil and therefore in the recycling of nutrients. Certain soil microorganisms such as mycorrhizal fungi can also increase the availability of mineral nutrients (*e.g.* phosphorus) to plants. Other soil microorganisms can increase the amount of nutrients present in the soil. For instance, nitrogen-fixing bacteria can transform nitrogen gas present in the soil atmosphere into soluble nitrogenous compounds that plant roots can utilize for growth. These microorganisms, which improve the fertility status of the soil and contribute to plant growth, have been termed 'biofertilizers' and are receiving increased attention for use as microbial inoculants in agriculture. Similarly, other soil microorganisms have been found to produce compounds (such as vitamins and plant hormones) that can improve plant health and contribute to higher crop yield. These microorganisms (called 'phytostimulators') are also studied for possible use as microbial inoculants to improve crop yield.

In contrast to these beneficial soil microorganisms, other soil microorganisms are pathogenic to plants and may cause considerable damage to crops. Large numbers of pathogenic microorganisms are routinely found in the soil and many of them can infect the plant through the roots. However, certain native microorganisms present in the soil are antagonistic to these pathogens and can prevent the infection of crop plants. Antagonism against plant pathogens usually involves competition for nutrients and/or production of inhibitory compounds such as secondary metabolites (antimicrobial metabolites and antibiotics) and extracellular enzymes. Other soil microorganisms produce compounds that stimulate the natural defense mechanisms of the plant and improve its resistance to pathogens. Collectively, these soil microorganisms have been termed as 'biopesticides' and represent an emerging and important alternative (*i.e.* biological control) to the use of chemical pesticides for the protection of crops against certain pathogens and pests.

Factors affecting microbial activities in soil

Several environmental conditions affect the density and composition of the microflora and frequently alter their activities in soil. Primary factors include soil moisture, aeration, temperature, pH, organic matter and inorganic fertilizers. Lesser variables are the secondary factors, which include crop rotation, season, soil depth, cultivation practices *etc.*

I. Primary factors

Soil moisture

Soil moisture is one of the important factors influencing the microbial population. Water is the major component of protoplasm; an adequate supply must be available for vegetative development of organisms. But, when it becomes excessive, proliferation is suppressed due to limitation in gaseous exchange and lowers the availability of O_2 supply creating anaerobic environment. Most of the organisms prefer a moisture percentage between 20 and 60 per cent. Many bacteria and fungi are able to adjust themselves to different moisture conditions. Under dry conditions, the bacteria may form spores that can resist the drought conditions. The fungi may sporulate or form chlamydospores to tide over adverse conditions. The protozoans also may form cysts and can survive under dry conditions. Actinomycetes are the chief group of organisms that prefer dry conditions. It is believed that the concentration of nutrients are diluted and also the aeration is very much limited at moisture and hence only the anaerobic and microaerophilic organisms can develop better.

Soil air

It is directly linked up the moisture level of the soil. Most of the forms are active in aerated soil and under waterlogged soils anaerobic and microaerophilic forms develop.

Soil temperature

It exhibits considerable influence on the microbial population. Though microorganisms have been found to exist under extreme temperature conditions, such as – 60° C and + 60° C the soil temperature usually does not reach such extremes. Microbial population varies both quantitatively and qualitatively under extreme conditions. In tropical and subtropical regions, temperatures vary widely in summer and winter seasons and the population may also be varied. In temperate regions, there is not much variation in temperatures of seasons and not much variation in the soil population. Soil temperature influences the temperature of air, water and solid phase of the soil. Soil water and temperature exert a combined influence on the microbial population.

Soil organic matter

Community size is related to the organic matter content, so that humus rich localities have the largest biological numbers. Organic matter content varied with soil types from less than 0.5 per cent in desert soils to 40 per cent in peaty soils. Organic matter being the chief source of energy and food for most soil organisms, it has great influence on the population. Nature of Organic matter is responsible for the differential stimulation of the population. There are several indirect effects of the organic matter on soil microflora. It influences the structure and texture of soil, besides enriching with nutrients for plants and microorganisms. Such influences on the soil also greatly affect the activity of the soil microorganisms.

Soil pH

It is a key factor influencing the microflora of soil. It influences enzyme system and thus plays an important role in microbial activity. In general, fungi thrice better than bacteria and actinomycetes in acid soils. Bacteria flourish well in neutral and alkaline soils. The saline and alkaline soils have different microflora. Salinity is due to excess salts; alkalinity is due to high H-ion concentration. Several direct and indirect effects of H-ion and salt concentration in soil are exerted on microbial population.

Fertilizer application

Application of fertilizers, to the soil improves the microbial activity because of the availability of more readily obtainable nutrients. Some fertilizers may however have inhibitory effect on specific bacterial types. Continuous application of ammoniacal fertilizers favours the growth of fungi due to the formation of nitric acid that inhibits the growth of bacteria and actinomycetes. Addition of nitrate inhibits the activity of free living N_2 fixing bacteria like Azotobacter. Some of the autotrophs are encouraged by the addition of fertilizers.

Cropping and vegetation

The crop plants exhibit two kinds of effects. One is through root exudates, which may have different compounds with reference to the crops grown. Vegetation has selective stimulation over population. Continuous cultivation leads to more microbial activity than the uncultivated land.

II. Secondary factors

Crop rotation

Crop rotation with different species like legumes, graminaceous plants, *etc.* brings about different stimulatory effect on the microflora. Some crops have deeper roots than others and some are more fibrous. Such variations bring about physical changes in soil, which in turn may have direct and indirect stimulatory effects on the soil microflora.

Cultural practices

Various cultural practices, such as tillage operations and irrigation have several physical and chemical changes in soil, which are reflected on the soil microflora. Through subsoil ploughing deeper layers may get better aeration and there may be quicker multiplication of aerobic organisms. Weeding, irrigation *etc.* may influence the microbial populations in the soil.

Soil depth

Most of the organisms are almost in top layers, largely in upper few centimeters and decline with greater depth, more active at few cm down and less active at deeper layers. Low O_2 and less sunlight in deeper layers reduce soil microbial population.

Season

Cell numbers are greatest during the spring and autumn and a decline occur during hot, dry and winter, the cells remain in a state of dormancy for biochemical inactivity. The influence is due to mainly the alterations in temperature, moisture as well as availability of organic matter.

III. Specific influences

- Sunlight favours algae and other autotrophs.
- Herbicide application has devasting effect on algae.
- Attack by neighbours – protozoa, nematode, earthworms consume algae.
- Presence of parasites and predators – *e.g.* viruses eat on specific bacteria
- Increase in bacterial population increases protozoans also.

IV. Other factors

Burning of topsoil

- Leads to partial sterilization of the top soil and may kill protozoan population, which may lead to the increase in bacterial population. This condition may affect biological equilibrium in soil.
- Application of nematicide, fungicide and bactericide may exhibit partial sterilization.Thus, every change in crop production directly or indirectly alters the soil microflora, still many more are not clearly understood.

Factors Affecting Distribution and Population of Soil Microorganisms

Soil microorganisms (Flora & Fauna), just like higher plants depends entirely on soil for their nutrition, growth and activity. The major soil factors which influence the microbial population, distribution and their activity in the soil are

1. Cultural practices
2. Soil fertility
3. Soil moisture
4. Soil temperature
5. Soil aeration
6. Light
7. Soil pH (H-ion Concentration)
8. Organic matter
9. Food and energy supply
10. Nature of soil
11. Microbial associations.

All these factors play a great role in determining not only the number and type of organism but also their activities. Variations in any one or more of these factors may lead to the changes in the activity of the organisms which ultimately affect the soil fertility level. Brief account of all these factors influencing soil micro flora / organisms and their activities is activities are discussed paragraphs.

1. Cultural practices (Tillage): Cultural practices *viz.* cultivation, crop rotation, application of manures and fertilizers, liming and gypsum application, pesticide/ fungicide and weedicide application have their effect on soil organism. Ploughing and tillage operations facilitate aeration in soil and exposure of soil to sunshine and thereby increase the biological activity of organisms, particularly of bacteria. Crop rotation with legume maintains the favorable microbial population balance, particularly of N2 fixing bacteria and thereby improve soil fertility.

- ❑ Liming of acid soils increases activity of bacteria and actinomycetes and lowers the fungal population. Fertilizers and manures applied to the soil for increased crop production, supply food and nutrition not only to the crops but also to microorganisms in soil and thereby proliferate the activity of microbes.
- ❑ Foliar or soil application of different chemicals (pesticides, fungicides, nematicides *etc.*) in agriculture are either degraded by the soil organisms or are liable to leave toxic residues in soil which are hazardous to cause profound reduction in the normal microbial activity in the soil.

2. Soil fertility: Fertility level of the soil has a great influence on the microbial population and their activity in soil. The availability of N, P and K required for plants as well as microbes in soil determines the fertility level of soil. On the other hand soil micro flora has greater influence on the soil fertility level.

6. Light: Direct sunlight is highly injurious to most of the microorganisms except algae. Therefore upper portion of the surface soil a centimeter or less is usually sterile or devoid of microorganisms. Effect of sunlight is due to heating and increase in temperature (More than 45°)

7. Soil Reaction / Soil pH: Soil reaction has a definite influence / effect on quantitative and qualitative composite on of soil microbes. Most of the soil bacteria, blue-green algae, diatoms and protozoa prefer a neutral or slightly alkaline reaction between PH 4.5 and 8.0 and fungi grow in acidic reaction between PH 4.5 and 6.5 while actinomycetes prefer slightly alkaline soil reactions. Soil reactions also influence the type of the bacteria present in soil. For example nitrifying bacteria (*Nitrosomonas & Nitrobacter*) and diazotrophs like *Azotobacter* are absent totally or inactive in acid soils, while diazotrophs like *Beijerinckia, Derxia,* and sulphur oxidizing bacteria like *Thiobacillus thiooxidans* are active in acidic soils.

8. Soil Organic Matter: The organic matter in soil being the chief source of energy and food for most of the soil organisms, it has great influence on the microbial population. Organic matter influence directly or indirectly on the population and activity of soil microorganisms. It influences the structure and texture of soil and thereby activity of the microorganisms.

9. Food and energy supply: Almost all microorganisms obtain their food and energy from the plant residues or organic matter / substances added to the soil. Energy is required for the metabolic activities of microorganisms. The heterotrophs utilize the energy liberated during the oxidation of complex organic compounds in soil, while autotrophs meet their energy requirement form oxidation of simple inorganic compounds (chemoautotroph) or from solar radiation (Photoautotroph). Thus, the source of food and energy rich material is essential for the microbial activity in soil. The organic matter, therefore serves both as a source of food nutrients as well as energy required by the soil organisms.

10. Nature of Soil: The physical, chemical and physico-chemical nature of soil and its nutrient status influence the microbial population both quantitatively and qualitatively. The chemical nature of soil has considerable effect on microbial population in soil. The soils in good physical condition have better aeration and moisture content which is essential for optimum microbial activity. Similarly nutrients (macro and micro) and organic constituents of humus are responsible for absence or presence of certain type of microorganisms and their activity. For example activity and presence of nitrogen fixing bacteria is greatly influenced by the availability of molybdenum and absence of available phosphate restricts the growth of *Azotobacter.*

11. Microbial associations / interactions: Microorganisms interact with each other giving rise to antagonistic or symbiotic interactions. The association existing between one organism and another whether of symbiotic or antagonistic influences the population and activity of soil microbes to a great extent. The predatory habit of protozoa and some mycobacteria which feed on bacteria may suppress or eliminate certain bacteria. On the other hand, the activities of some of the microorganisms are beneficial to each other. For instance organic acids liberated by fungi, increase in oxygen by the activity of algae, change in soil reaction *etc.* favors the activity or bacteria and other organisms in soil.

12. Root Exudates: In the soil where plants are growing the root exudates also affects the distribution, density and activity of soil microorganism. Root exudates and sloughed off material of root surfaces provide an abundant source of energy and nutrients and thus directly or indirectly influence the quality as well as quantity of microorganisms in the rhizosphere region. Root exudates contain sugars, organic acids, amino acids, sterols, vitamins and other growth factors which have the profound effect on soil microbes.

Methods for elucidating microbial biomass of soil

More recently newer methods have been developed that use measures of components of all cells to estimate biomass. These methods include:

a) Dilution plate counts and direct microscopical counting

Dilution plates usually only are able to culture between 1 and 10% of the viable organisms in soil samples. Direct microscopic observation methods (FITC, acridine orange staining, *etc.*) usually overestimate the number of cells because they include dead organism or other particles in their count.

b) Determination of cell wall components of bacteria

Bacteria contain specific cell wall components such as muramic acid that can be released by acid hydrolysis and analyzed by High Pressure Liquid Chromatography (HPLC). The amount of muramic acid in bacterial cell walls varies depending on whether the cells are gram positive or gram negative.

c) Soil fumigation method

The fumigation of soil with $CHCl_3$ followed by extraction with various salt solutions has been used to measure microbial biomass.

d) Determination of ATP content of soils

Adenosine tri phosphate is an important energy compound in the metabolism of all living organisms. ATP is considered to be a useful indicator of life in soil. ATP can be measured with great sensitivity. Though dependent on physiological state, ATP concentrations are fairly uniform relative to cell carbon for many bacteria, algae, and protozoa. ATP is an accepted indicator of microbial biomass in aquatic and terrestrial environments. Fireflies or lightening bugs make light within their bodies. This process is bioluminescence and is shared by many other organisms, mostly sea living organisms. The cells contain a chemical called luciferin and make an enzyme called luciferase. To make light, the luciferin combines with oxygen to form an inactive molecule called oxyluciferin. The wavelength of light given off is between 510 and 670 nm (pale reddish green color). Amount of light emitted in this reaction is directly proportional to the ATP concentration. Measurements of ATP can be used to estimate microbial biomass

4

Diversity of Uncultivable Microorganism – Metagenomic Approach

Metagenomics and its application in soil microbial community studies

The use of traditional microbiological culturing methods for the study of microbes has had limited success. It has been estimated that 99% of the microbes cannot be cultivated easily. Different problems have been faced by researchers during attempts to culture some microbes. Over the past decade, "metagenomics," which is culture-independent genomics analysis of microbes, has been developed to overcome these difficulties. Soil is considered to be a complex environment, which is a major reservoir of microbial genetic diversity. Metagenomic approaches promise increased accessibility of the genetic resources contained in the soil.

Metagenomics

Metagenomics was coined by Jo Handelsman in 2004. metagenomics is a new field of research that allows the study of genomes recovered from environmental samples. Metagenomics can be defined as "the application of modern genomics techniques to the study of communities of microbial organisms directly in their natural environments. The metagenomic methodology has been developed as an effective tool for the discovery of new natural products and microbial functions.

Applications of metagenomics

It provides the capacity to effectively characterize the genetic diversity present in water, soil and rumen source samples regardless of the availability of laboratory culturing techniques. Metagenomics is a new and exciting field of molecular biology that is likely to grow into a standard technique for understanding biological diversity.

Metagenomics technologies

Metagenomics is a means of systematically investigating, classifying and manipulating the entire genetic material isolated from environmental samples. The process involves isolating DNA from environmental sources and cloning it into vectors that replicate in cultured organisms.

Metagenomics process which relies on the efficiency of three main steps, *i.e.*

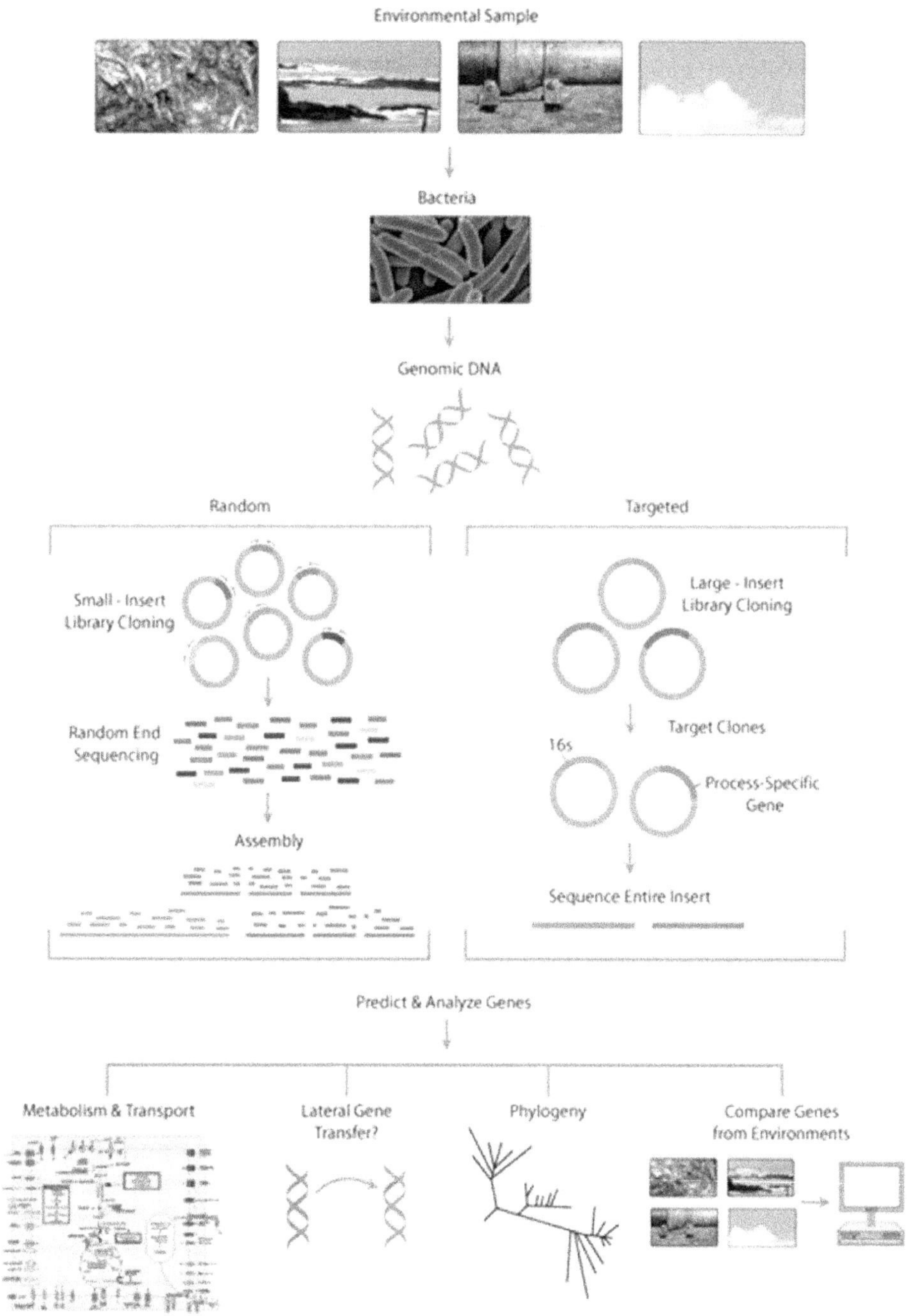

Steps involved in a metagenomics

1. Sampling and nucleic acids extraction
2. Library construction.
3. Analysis of metagenomic libraries

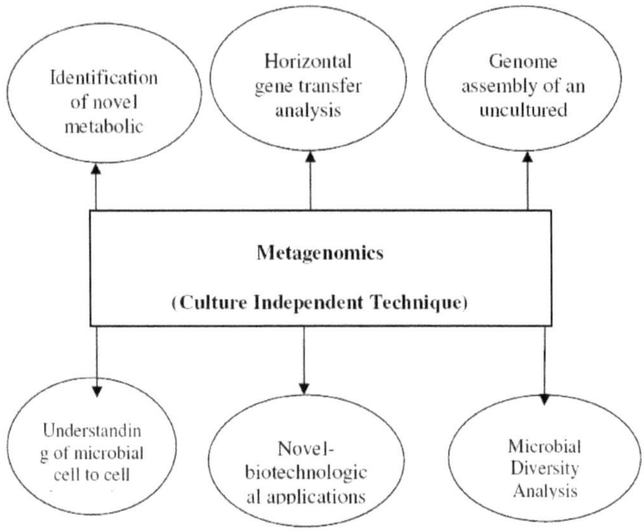

Applications of metagenomics.

Application of soil metagenomics

The development and application of metagenomics has enabled access to the uncultivated soil microbial community, availing a rich source of novel and useful biomolecules. Some examples of application of soil metagenomics are:

Target gene/natural product	Souce	Vector used for library construction	Host strain
Lipase, Esterase	Meadow Soil	PBluescript SK	*E. coli* DH5
Antibacterial, Hemolytic activities,Lipase,amylase,	North American Soil	PBeloBAC11	*E. coli* DH10B
nuclease 4-hydroxybutyrate dehydrogenase	Soil	PBluescript SK	*E. Coil* DH5
H+ antiporters	Soil	PBluescript SK	*E. coil* KNabc
Unique lipolytic activity	Forest Topsoil	Fosmid	*E.coil*
Novel biocatalysts	Unplanted field soil	Cosmid	*E.coil*
Alcohol oxidoreductase	Meadow, sugar beet field, cropland soil	PBluescript SK	*E.coil* DH5 alpha
Carbonyls formation	Soil	n.r	*E.coil*

Target gene/natural product	Souce	Vector used for library construction	Host strain
Coenzyme B (12)-dependent glycerol and diol dehydratases	Soil	PBluescript SK	*E.coil* DH5 alpha
esterase	Soil	pCCIOFS	*E.coil* EP-1300-TI
Lipase	Soil	pEpiFOS-5	*E.coil* EP1-100
Amidase	Soil and Enrichment Culture	PZero-2	*E.coil* TOP10
Amylase	Soil	Puc19	*E.coil* DH5 alpha
Polyketide synthase	Soil	Cosmid	*E.coil*
Aminoglycoside and tetracycline antibiotic	Soil	Pjn105	DH10B
resistance		pCF430	DH5 Alpha
Deoxybiolacein and broad spectrum antibiotic violacein	Soil	Cosmid	*E.coil*

Therefore, culture-independent methods are essential to understand the genetic diversity, population structure, and ecological roles of the majority of microorganisms. Metagenomics has potential to answer fundamental questions in microbial ecology, especially of soil ecosystems which contain the greatest diversity of microorganisms on earth, with 5,000–10,000 species of microorganism per gram of soil

More than 99% of bacteria from environmental samples remain 'unculturable' in the laboratory. Many of these 'unculturable' bacteria represent new phylotypes, families and divisions in domain bacteria and archaea. 'Unculturable' bacterial diversity presents a vast gene pool for biotechnological exploitation and poses a major challenge for microbiologists to understand their phylogenetic relationship and ecological significance. New culture techniques along with culture-independent methods, like PCR amplification from microbial community DNA (metagenome) and functional or sequence based screening of metagenomic DNA libraries are proving useful for exploitation of 'unculturable' bacteria. Environmental genome sequencing efforts are aiding better understanding of the community structure and their physiological importance in the ecosystem.

Definition

Metagenomics is the study of the collective genomes of the members of a microbial community. It involves cloning and analyzing the genomes without culturing the organisms in the community, thereby offering the opportunity to describe the planet's diverse microbial inhabitants, many of which cannot yet be cultured.

DNA is extracted directly from the community, cloned into a surrogate host, and then studied by sequencing or screening for expression of activities of interest. Many microbial communities have been tapped for metagenomic analyses. Following construction of a metagenomic library, two approaches can be taken to access the genomic information. Functional metagenomics requires that the host bacterium can express the recombinant DNA in either screens for active enzymes or antibiotic production or selections for growth under growth suppressive conditions (*e.g.*, nutrient deficiency or presence of an antibiotic). In sequence-based metagenomics, cloned DNA is randomly sequenced using vector-based primers or a specific gene is sought using complimentary oligonucleotides (oligos) to hybridize to arrayed metagenomic clones.

Culture-independent approach to assess the 'unculturable' bacterial diversity for novel biocatalysts and bioactive compounds

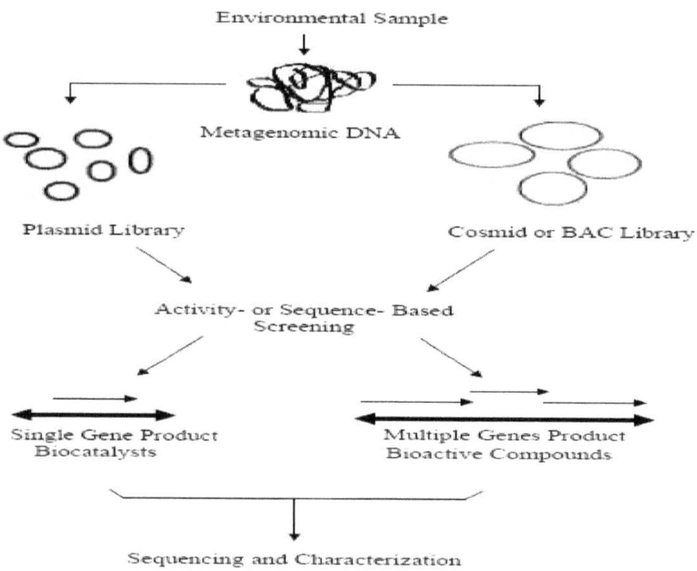

Strategies for the Metagenomic Analysis ofEnvironmental Microbial Communities

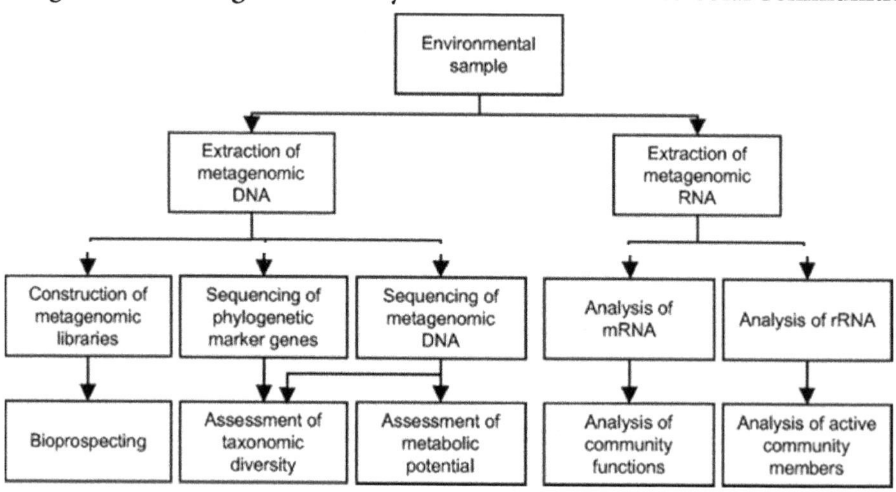

Metagenomics

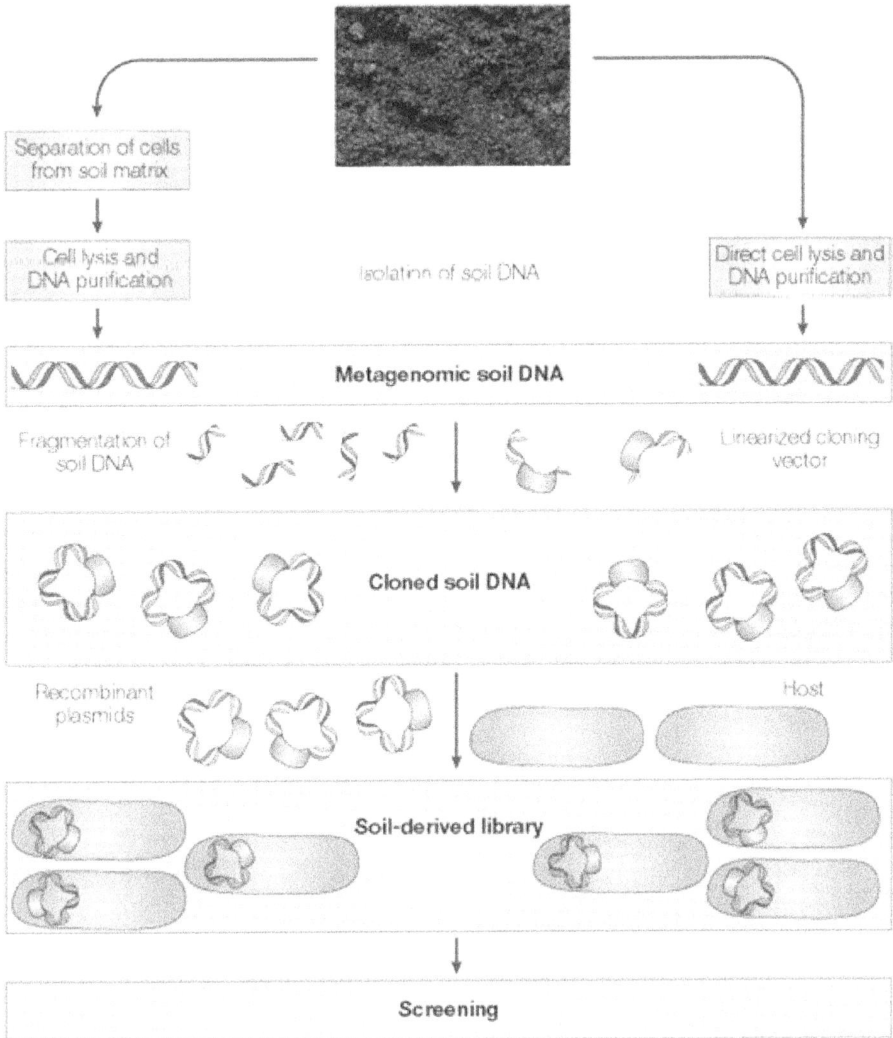

The rhizosphere is a key soil habitat, where the numerous interactions taking place between plant root and soil microorganisms will determine growth conditions for both the plant and the microorganisms in the rhizosphere. Indeed, the rhizosphere microbiota exerts an important influence on roots and on growth of the plant. Consequently, the rhizosphere represents an important soil habitat where introduction of beneficial microorganisms (*e.g.* biofertilizers, phytostimulators, and biopesticides) as inoculants can result in significant improvements in crop yield and /or crop quality.

Rhizosphere Concept

The root system of higher plants is associated not only with soil environment composed of inorganic and organic matter, but also with a vast community of metabolically active microorganisms. As living plants create a unique habitat around the roots, the microbial population on and around the roots is considerably higher than that of root free soil environment and the differences may be both quantitative and qualitative.

1. **Rhizosphere:** It is the zone/region of soil immediately surrounding the plant roots together with root surfaces, or it is the region where soil and plant roots make contact, or it is the soil region subjected to influence of plant roots and characterized by increased microbial.

2. **Rhizoplane:** Root surface along with the closely adhering soil particles is termed as rhizoplane.

Rhizosphere effect

It indicates the over all influence of plant roots on soil microorganisms. R: S ratio is used to find out the degree or extend of influence of the plant roots on soil microorganisms.

R:S ratio is defined as the ratio of microbial numbers per unit weight of rhizosphere (R), to the numbers in a unit weight of the adjacent non-rhizosphere soil (S). The rhizosphere effect is consistently greater for bacteria than for other microbes. The short, gram-negative rods, which almost invariably make up to a larger percent of the rhizosphere. On a generic basis, root effect appears to be more pronounced on Pseudomonas, Flavobacterium, Alcaligenes,Arthrobacter, Brevibacterium, Cornebacterium, Micrococcus, Xanthomonas, and Bacillus. Anaerobic bacteria are affected by the root; this may be attributed to the reduced O_2 tension resulting from root and microbial respiration. Bacterial density in the rhizosphere soil is 10^9 per gram and it may cover 4-10 % of the root area. Fast growers and metabolic active population are dominating in the rhizosphere region. This suggests that the rhizosphere flora has a greater ability to effect rapid biochemical changes than the organisms of fallow land. Selective influence of he root system is exhibited in few cases like preferential stimulation of G-ve, non spore forming short rods, vitamin requiring bacteria, abundance of N_2 fixing and P solubilizing organisms *etc*., The spectrum of genera varies with plant species, age and kind of the soil. Actinomycetes and algae are not significantly benefited by their proximity to roots.

Root exudates

One of the most important factors responsible for rhizosphere effect is the great variety of organic substances available at the root region by the way of exudates from roots, which directly or indirectly influence the quality and quantity of the micro organism in the root region. The substances exuded by plant roots include amino acids (asparagine,alanine, leucine, serine, glycine, lysin, prolein, trytophane), sugars (glucose, fructose, sucrose, xylose, arabinose, oligosaccharides), organic acids (tartaric, oxalic, malic, acetic, propionic, succinic, fumaric, butric), growth factors (biotin, thiamine, niacin, inositol, pyridoxin, p- amino benzoic acid), nucleotides, flavanones, enzymes and many other unidentified substances.

The nature and amount of substances exuded are dependent on the species of plant, age and environmental conditions under they grow. The products of photosynthesis are translocated to the root system in less than 12 hours and indicating the influence of metabolism of plants determining the rhizosphere effect. The root cap and ares of active growth are primary regions of root exudation.

Influence of the rhizosphere microflora on plant growth

The rhizosphere community has either a favourable or detrimental influence of plant development. Because the microflora is so intimately related with the root system, partially covering its surface, any beneficial or toxic substance produced can cause an immediate response. Vast microorganisms demand a variety of anions and cations for their own development and as a result immobilization take place. Aerobic organisms remove O_2 from the environment and add CO_2 and due to lowering of O_2 or increasing of CO_2 tension, it may reduce root elongation, development and reduce the water and nutrient uptake. It may however favour plant development by producing growth stimulating substances, forming of stable soil structure, releasing elements in organic forms through the mineralization of organic complexes, and by entering into symbiotic root associations.

Influence of the Plant on microflora

The microflora is affected in many ways by the growing plant. The important contribution is the provision of excretion products and sloughed off tissues to serve as energy, carbon, nitrogen or growth factors. Due to the large quantities of CO_2 liberated by the root zone during respiration, the carbonic acid is formed. It solubilizes the unavailable elements to make available and that will be assimilated by the microflora.

Historical Background

Term "Rhizosphere" was introduced for the first time by the German scientist **Hiltner(1904)** to denote that region of soil which is subjected to the influence of plant roots. The concept of "Rhizosphere Phenomenon" which shows the mutual interaction of roots and microorganisms was came into existence with the work of **Starkey et al (1929), Clark (1939) and Rauath and Katznelson (1957).**

N. V. Krassinikov (1934)found that free\ living nitrogen fixing bacteria, *Azotobacter*were unable to grow in the wheat rhizosphere.

Starkey (1938) examined the rhizosphere region of some plant species and demonstrated the effect of root exudates on the predominance of bacterial population in particular and other soil microorganisms in general in the rhizosphere region. Thus, he put forth the concept of **"Rhizosphere effect / phenomenon"** for the first time.

F E Clark (1949) introduced / coined the term **"Rhizoplane"** to denote the root surface together with the closely adhering soil particles.

R. I. Perotti (1925) suggested the boundaries of the rhizosphere region and showed that it was bounded on one side by the general soil region (called as Edaphosphere) and on the other side by the root tissues (called Histosphere).

G. Graf and S. Poschenrieder (1930) divided the rhizosphere region into two general areas *i.e.* outer rhizosphere and inner rhizosphere for the purpose of describing the same site of microbial action.

H. Katznelson (1946) suggested the **R:S ratio** *i.e.* the ratio between the microbial population in the rhizosphere (R) and in the soil (S) to find out the degree or extent of plant roots effect on soil microorganisms. R: S ratio gives a good picture of the relative stimulation of the microorganisms in the rhizosphere of different plant species.

R: S ratio is defined as the ratio of microbial population per unit weight of rhizosphere soil (R), to the microbial population per unit weight of the adjacent non-rhizosphere soil (S)

A. G. Lochhead and H. Katznelson (1940) examined in detail the qualitative differences between the microflora of the rhizosphere and microflora of the non-rhizosphere region and reported that gram-negative, rod shaped and non-spore forming bacteria are abundant in the rhizosphere than in the non-rhizosphere soil

C. Thom and H. Humfeld (1932) found that corn roots in acidic soils yielded predominantly *Trichoderma* while roots from alkaline soils mainly contained *Penicillium.*

M J. Timonin (1940) reported some differences in the fungal types and population in the rhizosphere of cereals and legumes. R: S ratio of fungal population was believed to be narrow in most of the plant species, usually not exceeding 10.

E. A. Peterson and others (1958) reported that the plant age and soil type influence the nature of fungal flora in the rhizosphere, and the number of fungal population gradually increases with the age of plant.

M. Adati (1932) studied many crops and found that though actinomycetes were relatively less stimulated than bacteria, but in some cases the R: S ratio of actinomycetes was as high as 62.

R. Venkatesan and G. Rangaswami (1965) studied the rhizosphere effect in rice plant on bacteria, actinomycetes and fungi and reported that (i) for actinornycetes R: S was more (ranging from 0 to 25) depending on the age of plant roots and the dominant genera reported were Nocardia, (ii) R:S ratio reduced with the depth of soil.

E. A. Gonsalves and V. S. Yalavigi (1960) reported the presence of greater number of algae in the rhizosphere

J. W. Rouatt et al reported positive rhizosphere effect on protozoa, but a negative effect on algae in wheat plants.

Microorganisms in the Rhizosphere and Rhizosphere Effect

The rhizosphere region is a highly favorable habitat for the proliferation, activity and metabolism of numerous microorganisms. The rhizosphere microflora can be enumerated intensively by microscopic, cultural and biochemical techniques. Microscopic techniques reveal the types of organisms present and their physical association with the outer root tissue surface / root hairs. The cultural technique most commonly followed is "serial dilution and plate count method" which reveal the quantitative and qualitative population of microflora. At the same time, a cultural method shows the selective enhancement of certain categories of bacteria. The biochemical techniques used are designed to measure a specific change brought about by the plant or by the microflora. The rhizosphere effect on most commonly found microorganisms *viz.* bacteria, actinomycetes, fungi, algae and protozoa is being discussed herewith in the following paragraphs.

A. Bacteria

The greater rhizosphere effect is observed with bacteria (R: S values ranging from 10-20 or more) than with actinomycetes and fungi. Gram-negative, rod shaped, non-sporulating bacteria which respond to root exudates are predominant in the rhizosphere *(Pseudomonas, Agrobacterium).* While Gram-positive, rods, Cocci and aerobic spore forming *(Bacillus, Clostridium)* are comparatively rare in the rhizosphere. The most common genera of bacteria are: *Pseudomonas, Arthrobacter, Agrobacterium, Alcaligenes, Azotobacter, Mycobacterium, Flavobacter, Cellulomonas, Micrococcus* and others have been reported to be either abundant or sparse in the rhizosphere. From the agronomic point of view, the abundance of nitrogen fixing and phosphate solubilizing bacteria in the rhizosphere assumes a great importance. The aerobic bacteria are relatively less in the rhizosphere because of the reduced oxygen levels due to root respiration. The bacterial population in the rhizosphere is enormous in the ranging form 10^8 to 10^9 per gram of rhizosphere soil. They cover about 4-10% of the total root area occurring profusely on the root hair region and rarely in the root tips. There is predominance of amino acids and growth factors required by bacteria, are readily provided by the root exudates in the region of rhizosphere.

B. Fungi

In contrast to their effects on bacteria, plant roots do not alter / enhance the total count of fungi in the rhizosphere. However, rhizosphere effect is selective and significant on specific fungal genera *(Fusarium, Verticillium, Aspergillus* and *Penicillium)* which are stimulated. The R:S ratio of fungal population is believed to be narrow in most of the plants, usually not exceeding to 10. The soil / serial dilution and plating technique used for the enumeration of rhizosphere fungi may often give erratic results as most of the spore formers produce abundant colonies in culture media giving a wrong picture / estimate *(eg Aspergilli* and *Penicillia).* In fact the mycelial forms are more dominant in the field. The zoospore / forming lower fungi such as *Phytophthora, Pythium, Aphanomyces* are strongly attracted to the roots in response to particular chemical compounds excreted by the roots and cause diseases under favorable conditions. Several fungi *eg Gibberella* and *fujikurio* produces *phytohormones* and influence the plant growth.

C. Actinomycetes, Protozoa and Algae

Stimulation of actinomycetes in the rhizosphere has not been studied in much detail so far. It is generally understood that the actinomycetes are less stimulated in the rhizosphere than bacteria. However, when antagonistic actinomycetes increase in number they suppress bacteria. Actinomycetes may also increase in number when antibacterial agents are sprayed on the crop. Among the actinomycete, the phosphate solublizers (eg. *Nocardia, Streptomyces)* have a dominant role to play.

As rule actinomycetes, protozoa and algae are not significantly influenced by their proximity to the plant roots and their R: S ratios rarely exceed 2 to 3: 1 and around roots of plants, R: S ratio for these microorganisms may go to high. Because of large bacterial community, an increase in the number or activity of protozoa is expected in the rhizosphere. Flagellates and amoebae are dominant and ciliates are rare in the region.

Factors affecting microbial flora of the Rhizosphere / Rhizosphere Effect

The most important factors which affect / influence the microbial flora of the rhizosphere or rhizosphere effect are: soil type & its moisture, soil amendments, soil PH, proximity of root with soil, plant species, and age of plant and root exudates.

A. Soil type and its moisture: In general, microbial activity and population is high in the rhizosphere region of the plants grown in sandy soils and least in the high humus soils, and rhizosphere organisms are more when the soil moisture is low. Thus, the rhizosphere effect is more in the sandy soils with low moisture content.

B. Soil amendments and fertilizers: Crop residues, animal manure and chemical fertilizers applied to the soil cause no appreciable effect on the quantitative or qualitative differences in the microflora of rhizosphere. In general, the character of vegetation is more important than the fertility level of the soil.

C. Soil PH/ Rhizosphere PH: Respiration by the rhizosphere microflora may lead to the change in soil rhizosphere PH. If the activity and population of the rhizosphere microflora is more, then the PH of rhizosphere region is lower than that of surrounding soil or non-rhizosphere soil. Rhizosphere effect for bacteria and protozoa is more in slightly alkaline soil and for that of fungi is more in acidic soils.

D. Proximity of root with Soil: Soil samples taken progressively closer to the root system have increasingly greater population of bacteria, and actinomycetes and decreases with the distance and depth from the root system. Rhizosphere effect decline sharply with increasing distance between plant root and soil.

E. Plant Species: Different plant species inhabit often some what variable microflora in the rhizosphere region. The qualitative and quantitative differences are attributed to variations in the rooting habits, tissue composition and excretion products. In general, legumes show / produce a more pronounced rhizosphere effect than grasses or cereals. Biennials, due to their long growth period exert more prolonged stimulation on rhizosphere effect than annuals.

F. Age of Plant: The age of plant also alter the rhizosphere microflora and the stage of plant maturity controls the magnitude of rhizosphere effect and degree of response to specific microorganisms. The rhizosphere microflora increases in number with the age of the plant and reaching at peak during flowering which is the most active period of plant growth and metabolism. Hence, the rhizosphere effect was found to be more at the time of flowering than in the seedling or full maturity stage of the plants. The fungal flora (especially, Cellulolytic and Amylolytic) of the rhizosphere usually increases even after fruiting and the onset of senescence due to accumulation of moribund tissue and sloughed off root parts / tissues: whereas, bacterial flora of the rhizosphere decreases after the flowering period and fruit setting.

G. Root / exudates /excretion: One of the most important factors responsible for rhizosphere effect is the availability of a great variety of organic substances at the root region by way of root exudates/excretions. The quantitative and qualitative differences in the microflora of the rhizosphere from that of general soil are mainly due to influences of root exudates. The spectrum of chemical composition root exudates varies widely, and hence their influence on the microflora also varies widely.

Root exudates are composed of the chemical substances like

Sr. No	Root Executes	Chemical Substances
1	Amino Acids	All naturally occurring amino acids.
2	Organic acids	Acetic, butyric, citric, fumaric, lactic, malic, propionic, succinic *etc.*
3	Carbohydrates / sugars	Arabinose, fructose, galactose, glucose, maltose, mannose, oligosaccharides, raffinose, ribose, sucrose, xylose *etc.*
4	Nucleic acid derivatives	Adenine, cystidine, guanine, undine
5	Growth factors (phytohormones)	Biotin, choline, inositol, pyridoxine *etc*
6	Vitamins	Thiamine, nicotinic acid, biotin *etc*
7	Enzymes	Amylase, invertase, protease, phosphatase *etc.*
8	Other compounds	Auxins, glutamine, glycosides, hydrocyanic acid peptides, Uv-absorbing compounds, nematode attracting factors, spore germination stimulators, spore inhibitors *etc.*

The nature and amount of chemical substances thus exuded are dependent on the species of plant, plant age, inorganic nutrients, and temperature, light intercity, O2 / CO_2 level, root injury *etc*. Another source of nutrients for the microorganisms in the rhizosphere region is the sloughed off root epidermis which exert selective stimulation effect on some specific groups of microorganisms. For instance, glucose and amino acids in the exudates readily attract Gram-negative rods which predominantly colonize the roots. Sugars and amino acids in the root exudates stimulate the germination of chlamydospores and other resting spores of fungi; stimulation effect of root exudates on plant pathogenic fungi, nematodes is also well known.

Alterations in Rhizosphere Microflora

Foliar application of various chemicals leads to alterations in the rhizosphere microflora by changing the pattern of root exudates. The pattern of the rhizosphere microflora *i.e.* numbers and species composition can be changed / altered by various factors, such as: (i) Soil amendments, (ii) Foliar application of fertilizers / nutrients, fungicides, insecticides and hormones and (iii) Bacterization / microbial seed inoculants.

A. Soil amendments

Soil amendments with inorganic and organic fertilizers can alter the rhizosphere microflora and an understanding of the type of changes in the microflora can be useful in the indirect control of pathogens. Dwivedi and Chaube (1985) showed that amendment of soil with neem-cake can stimulate the activity of actinomycetes which results into the reduction of propagules of *Macrophomina phaseolina.* It is also known to control phytopathogenic nematodes in soil by stimulating nematode trapping fungi. Amendment of soil with castor and bean leaves stimulate the activity of *Trichoderma viride* and *Penicillium* in the rhizosphere leading to the control of *Sclerotium rolfsii.*

B. Foliar application of fertilizers and agrochemicals

Translocation of photosynthete from leaves to roots takes place as a part of the normal metabolic activity in plants. Therefore, organic substances, including plant protection chemicals (fungicides, insecticides), growth regulators and plant nutrients applied to foliage / leaves get absorbed into the leaf tissue and further get translocated to roots along with photosynthates. Many workers have reported that foliar application with various chemicals cause marked alterations in the number and kind / qualities of microorganisms in the rhizosphere of several cereals and leguminous crop plants. Thus, such an approach can be used as a new tool in the biological control of root diseases, stimulation of activity of nitrogen-fixing bacteria and other beneficial microorganisms in the soil.

C. Seed treatment with bio inoculants

Bio inoculants such as *Azotobacter, Beijerinckia, Azospirillum, Rhizobium or* P -solubilizing microorganisms (*e.g. Bcillus, plymyxa, Azotobacter croococcum, Aspergillus niger, Penicillium digitatum etc.*) When applied to the seed / soil helps in the establishment of beneficial microorganisms in the rhizosphere region which will further benefit in plant growth, encourage inhibition of plant pathogenic organisms in the root vicinity and enrich the soil with added microbial bio-mass.

Associative and Antagonistic activities in the Rhizosphere

In natural environments (*e.g.* Soil, Air, Water *etc.*) a number of relationships exist between individual microbes, microbial species and between individual cells.

The composition of microflora of any habitat (soil / rhizosphere) is governed by the biological equilibrium created by the associations and interactions of all individuals found in the community. In soil and rhizosphere region, many microorganisms live in close proximity and their interactions with each other may be associative or antagonistic.

A. Associative interactions / activities in rhizosphere: The dependence of one microorganism upon another for extra-cellular products (*e.g.* amino acids & growth promoting substances) can be regarded as an associative activity / effect in rhizosphere. There is an increase in the exudation of amino acids, organic acids and monosaccharide by plant roots in the presence of microorganisms. Gibberellins and gibberellin- like substances are known to be produced by bacterial genera *viz Azotobacter, Arthrobacter, Pseudomonas,* and *Agrobacterium* which are commonly found in the rhizosphere. Microorganisms also influence root hair development, mucilage secretion and lateral root development. Fungi inhabiting the root surface facilitate the absorption of nutrient by the roots.

Mycorrhiza is one of the best known associative / symbiotic interactions which exist between the roots of higher plants and fungi. This mycorrhizal association has been found to improve plant growth through better uptake of phosphorus and zinc from soil, suppression of root pathogenic fungi and nematodes. Another example is association between the bacterium *Rhizobium* and roots of legumes and *Azospirillum* with cereal crops (wheat, rye, bajara, maize *etc*).

B. Antagonistic interactions / activities in rhizosphere: The biochemical qualities of root exudates and the presence of antagonistic microorganisms, plays important role in encouraging or inhibiting the soil borne plant pathogens in the rhizosphere region. Several mutualistic, communalistic, competitive and antagonistic interactions exist in the rhizosphere. The number and qualities of antagonistic microorganisms in the rhizosphere could be increased through artificial means such as fertilizer application, organic amendments, foliar spraying of chemicals *etc.*

Antagonistic microorganisms in the rhizosphere play an important role in controlling some of the soil borne plant pathogens. Stanier et al (1966) discovered the bacterial strain *Pseudomonas fluorescens* and the fluorescent pigments of this species in biological control of root pathogens. Strains of *P. fluorescence* are collectively called as **"Fluorescent Pseudomonads"**. They produce variety of biologically active compounds such as plant growth substances, cyanides, antibiotics and iron chelating substances called **"Siderophores"** Rovira and Campbell (1975), showed that bacterial strains of *P fluorescens* could lyse the hyphae of *Gaumannomyces graminis var. Tritici,* the causative agent of take-all disease of wheat. Fluorescent pseudomonads *(P. fluorescens, P. putida)* are known to produce iron chelating substances called Siderophores. These are low molecular weight, extra cellular, iron-binding agents produced by pseudomonads in response to low iron stress or when Fe3 is in short supply. Thus, iron stress triggers the formation of iron-binding ligands called siderophores. Siderophores contains the pigments Pyovirdin (Fluorescent) and Pyocyanin (non-Fluorescent) having iron chelating properties. Another pigment "Pseudobactin" is a fluorescent chelator of iron which is known to promote plant growth and inhibition of pathogenic bacteria in the rhizosphere.

C. Changes in morphology and physiology of host plant: Changes in the physiology and morphology of host plant influence the rhizosphere microflora through root exudations. Hence, significant changes in the rhizosphere microflora of diseased plants were reported which are attributed to the nature and severity of the disease. Systemic virus diseases cause marked changes in the plant morphology and physiology to drastically alter the rhizosphere microflora.

D. Increase in antagonists activity: Root exudates provide a food base for the growth of antagonistic organisms which plays an important role in controlling / suppressing some of the soil borne plant pathogens. Generally, rhizosphere of the resistant plant varieties harboure moer number of *Streptomyces* and *Trichoderma* than that of susceptible varieties. For example in the rhizosphere of pigeon pea varieties resistant to *Fusarium udum,* the population of *Streptomyces* was found more which inhibited the growth of the pathogen. High density of *Trichoderma viride* in the rhizosphere of Tomato varieties resistant to *Verticillium* wilt has been reported with its ability to reduce the severity of wilt in susceptible plants.

E. Inhibition of pathogen: Root exudates containing toxic substances such as glycosides and hydrocyanic acid may inhibit the growth of pathogens in the rhizosphere. It has been reported that root exudates from resistant varieties of Flax (*e.g.* Bison) excrete a glucoside which on hydrolysis produces hydrocyanic acid that inhibits Fusarium oxysporum, the flax root pathogen. Exudates of resistant pea reduce the germination of spores of Fusarium oxysporum.

In this light, the rhizosphere may be considered as a microbiological buffer zone in which the microflora serves to protect the plants against the attack of the pathogens.

F. Attraction of bacteria and nematodes: Root exudates attracts phytopathogenic bacteria and fungi in the rhizosphere for example *Agrobacterium tumefaciens* have been reported to be attracted to the roots of the host plants like peas, maize, onion, tobacco, tomato and cucumber.

Host root exudates also influence phytopathogenic nematodes in two ways: (i) though stimulation of egg-hatching process and (ii) attraction of larvae towards plant roots.

6

Plant Growth Promoting Rhizobacteria (PGPR)

Rhizobacteria are root-colonizing bacteria that form symbiotic relationships with many plants. The name comes from the Greek rhiza, meaning root. Though parasitic varieties of rhizobacteria exist, the term usually refers to bacteria that form a relationship beneficial for both parties (mutualism). They are an important group of microorganisms used in biofertilizer. Biofertilization accounts for about 65% of the nitrogen supply to crops worldwide. Rhizobacteria are often referred to as plant growth-promoting rhizobacteria, or PGPRs. The term PGPRs was first used by Joseph W. Kloepper in the late 1970s and has become commonly used in scientific literature. PGPRs have different relationships with different species of host plants. The two major classes of relationships are rhizospheric and endophytic. Rhizospheric relationships consist of the PGPRs that colonize the surface of the root, or superficial intercellular spaces of the host plant, often forming root nodules. The dominant species found in the rhizosphere is a microbe from the genus Azospirillum. Endophytic relationships involve the PGPRs residing and growing within the host plant in the apoplastic space.

Nitrogen fixation

Nitrogen fixation is one of the most beneficial processes performed by rhizobacteria. Nitrogen is a vital nutrient to plants and gaseous nitrogen (N_2) is not available to them due to the high energy required to break the triple bonds between the two atoms. Rhizobacteria, through nitrogen fixation, are able to convert gaseous nitrogen (N_2) to ammonia (NH_3) making it an available nutrient to the host plant which can support and enhance plant growth. The host plant provides the bacteria with amino acids so they do not need to assimilate ammonia. The amino acids are then shuttled back to the plant with newly fixed nitrogen. Nitrogenase is an enzyme involved in nitrogen fixation and requires anaerobic conditions. Membranes within root nodules are able to provide these conditions. The rhizobacteria require oxygen to metabolize, so oxygen is provided by a hemoglobin protein called leghemoglobin which is produced within the nodules. Legumes are well-known nitrogen-fixing crops and have been used for centuries in crop rotation to maintain the health of the soil.

Symbiotic relationships

The symbiotic relationship between rhizobacteria and their host plants is not without costs. For the plant to be able to benefit from the added available nutrients provided by the rhizobacteria, it needs to provide a place and the proper conditions for the rhizobacteria to live. Creating and maintaining root nodules for rhizobacteria can cost between 12–25% of the plant's total photosynthetic output. Legumes are often able to colonize early successional environments due to the unavailability of nutrients. Once colonized, though, the rhizobacteria make the soil surrounding the plant more nutrient rich, which in turn can lead to competition with other plants. The symbiotic relationship, in short, can lead to increased competition.

PGPRs increase the availability of nutrients through the solubilization of unavailable forms of nutrients and by the production of siderophores which aids in the facilitating of nutrient transport. Phosphorus, a limiting nutrient for plant growth, can be plentiful in soil, but is most commonly found in insoluble forms. Organic acids and phosphotases released by rhizobacteria found in plant rhizospheres facilitate the conversion of insoluble forms of phosphorus to plant-available forms such as $H_2PO_4^-$. PGPR bacteria include Pseudomonas putida, Azospirillum fluorescens, and Azospirillum lipoferum and notable nitrogen-fixing bacteria associated with legumes includes Allorhizobium, Azorhizobium, Bradyrhizobium, and Rhizobium.

Though microbial inoculants can be beneficial for crops, they are not widely used in industrial agriculture, as large-scale application techniques have yet to become economically viable. A notable exception is the use of rhizobial inoculants for legumes such as peas. Inoculation with PGPRs ensures efficient nitrogen fixation, and they have been employed in North American agriculture for over 100 years.

Plant growth-promoting rhizobacteria

Plant growth-promoting rhizobacteria (PGPR) were first defined by Kloepper and Schroth to describe soil bacteria that colonize the roots of plants following inoculation onto seed and that enhance plant growth. The following are implicit in the colonization process: ability to survive inoculation onto seed, to multiply in the spermosphere (region surrounding the seed) in response to seed exudates, to attach to the root surface, and to colonize the developing root system. The ineffectiveness of PGPR in the field has often been attributed to their inability to colonize plant roots. A variety of bacterial traits and specific genes contribute to this process, but only a few have been identified. These include motility, chemotaxis to seed and root exudates, production of pili or fimbriae, production of specific cell surface components, ability to use specific components of root exudates, protein secretion, and quorum sensing. The generation of mutants altered in expression of these traits is aiding our understanding of the precise role each one plays in the colonization process.

Progress in the identification of new, previously uncharacterized genes is being made using nonbiased screening strategies that rely on gene fusion technologies. These strategies employ reporter transposons and in vitro expression technology (IVET) to detect genes expressed during colonization.

Using molecular markers such as green fluorescent protein or fluorescent antibodies, it is possible to monitor the location of individual rhizobacteria on the root using confocal laser scanning microscopy. This approach has also been combined with an rRNA-targeting probe to monitor the metabolic activity of a rhizobacterial strain in the rhizosphere and showed that bacteria located at the root tip were most active.

Mechanisms of action

PGPRs enhance plant growth by direct and indirect means, but the specific mechanisms involved have not all been well-characterized, Direct mechanisms of plant growth promotion by PGPRs can be demonstrated in the absence of plant pathogens or other rhizosphere microorganisms, while indirect mechanisms involve the ability of PGPRs to reduce the harmful effects of plant pathogens on crop yield. PGPRs have been reported to directly enhance plant growth by a variety of mechanisms: fixation of atmospheric nitrogen transferred to the plant, production of siderophores that chelate iron and make it available to the plant root, solubilization of minerals such as phosphorus, and synthesis of phytohormones. Direct enhancement of mineral uptake due to increases in specific ion fluxes at the root surface in the presence of PGPRs has also been reported. PGPR strains may use one or more of these mechanisms in the rhizosphere. Molecular approaches using microbial and plant mutants altered in their ability to synthesize or respond to specific phytohormones have increased understanding of the role of phytohormone synthesis as a direct mechanism of plant growth enhancement by PGPRs. PGPR that synthesize auxins and cytokinins or that interfere with plant ethylene synthesis have been identified.

Pathogenic roles

Studies conducted on sugar beet crops found that some root-colonizing bacteria were deleterious rhizobacteria (DRB). Sugar beet seeds inoculated with DRB had reduced germination rates, root lesions, reduced root elongation, root distortions, increased fungi infection, and decreased plant growth. In one trial the sugar beet yield was reduced by 48%.

Six strains of rhizobacteria have been identified as being DRB. The strains are in the genera Enterobacter, Klebsiella, Citrobacter, Flavobacterium, Achromobacter, and Arthrobacter. Due to a large number of taxonomic species yet to be described, complete characterization has not been possible as DRB are highly variable.

The presence of PGPRs has proven to reduce and inhibit the colonization of DRB on sugar beet roots. Plots inoculated with PGPRs and DRBs had an increase in production of 39% while plots only treated with DRBs had a reduction in production of 30%.

Biocontrol

Rhizobacteria are also able to control plant diseases that are caused by other bacteria and fungi. Disease is suppressed through induced systematic resistance and through the production of antifungal metabolites. Pseudomonas biocontrol strains have been genetically modified to improve plant growth and improve the disease resistance of agricultural crops. In agriculture, inoculant bacteria are often applied to the seed coat of seeds prior to being sown. Inoculated seeds are more likely to establish large enough rhizobacterial populations within the rhizosphere to produce notable beneficial effects on the crop.

The group of bacteria that colonize roots or rhizosphere soil and beneficial to crops are referred to as plant growth promoting rhizobacteria (PGPR). The PGPR inoculants currently commercialized that seem to promote growth through at least one mechanism; suppression of plant disease (termed Bioprotectants), improved nutrient acquisition (termed Biofertilizers), or phytohormone production (termed Biostimulants).

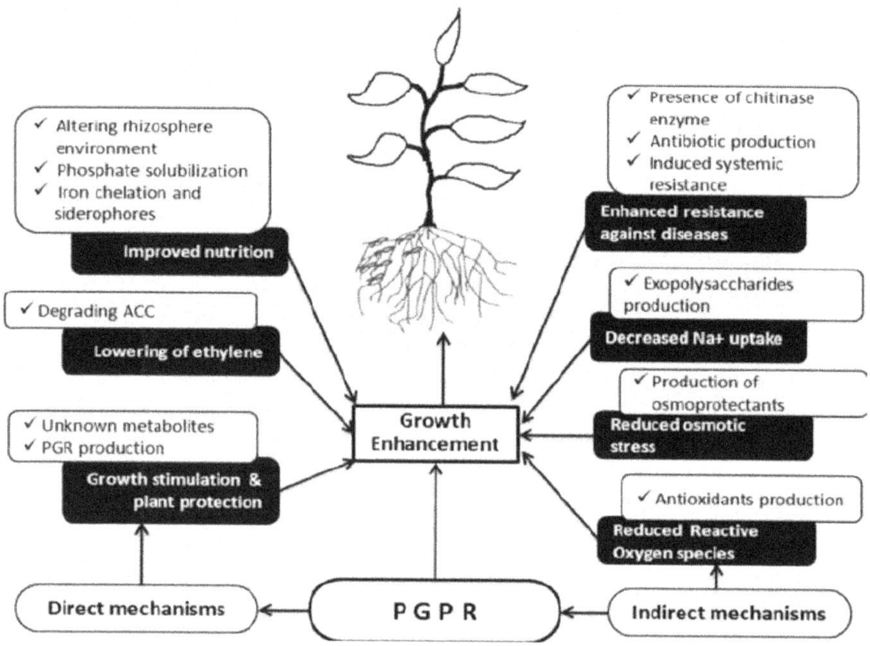

Species of Pseudomonas and Bacillus can produce as yet not well characterized phytohormones or growth regulators that cause crops to have greater amounts of fine roots which have the effect of increasing the absorptive surface of plant roots for uptake of water and nutrients. These PGPR are referred to as Biostimulants and the phytohormones they produce include indole-acetic acid, cytokinins, gibberellins and inhibitors of ethylene production.

Recent advances in molecular techniques also are encouraging in that tools are becoming available to determine the mechanism by which crop performance is improved using PGPR and track survival and activity of PGPR organisms in soil and roots. The science of PGPR is at the stage where genetically modified PGPR can be produced. PGPR with antibiotic, phytohormone and siderophore production can be made.

Despite of promising results, biofertilizers has not got widespread application in agriculture mainly because of the variable response of plant species or genotypes to inoculation depending on the bacterial strain used. Differential rhizosphere effect of crops in harbouring a target strain or even the modulation of the bacterial nitrogen fixing and phosphate solubilizing capacity by specific root exudates may account for the observed differences. On the other hand, good competitive ability and high saprophytic competence are the major factors determining the success of a bacterial strain as an inoculant.

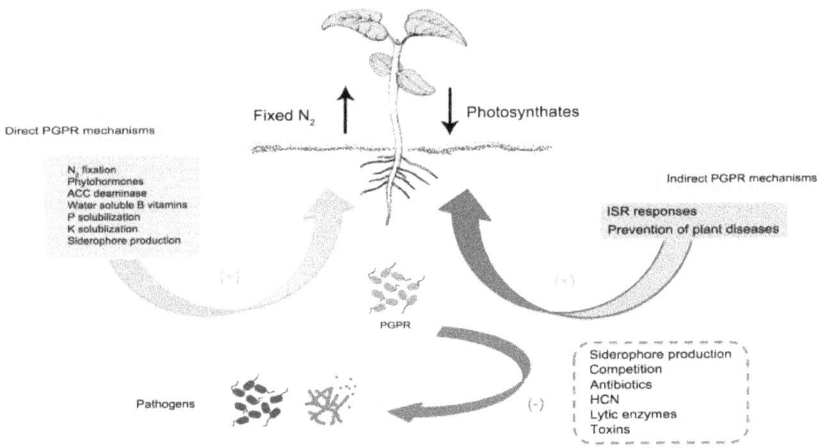

Studies to know the synergistic activities and persistence of specific microbial populations in complex environments, such as the rhizosphere, should be addressed in order to obtain efficient inoculants.

Phyllosphere Bacteria

Aerial plant surfaces represent the largest biological interface on Earth and provide essential services as sites of carbon dioxide fixation, molecular oxygen release, and primary biomass production. Rather than existing as axenic organisms, plants are colonized by microorganisms that affect both their health and growth.

For terrestrial plants, the phyllosphere represents the interface between the above-ground parts of plants and the air. Conservative estimates indicate that the roughly 1 billion square kilometers of worldwide leaf surfaces host more than 10^{26} bacteria, which are the most abundant colonizers of this habitat. The overall microbiota in this ecosystem is thus sufficiently large to have an impact on the global carbon and nitrogen cycles. Additionally, the phyllosphere inhabitants influence their hosts at the level of the individual plants. To a large extent, interest in phyllosphere microbiology has been driven by investigations on plant pathogens. Their spread, colonization, survival, and pathogenicity mechanisms have been the subject of numerous studies. Much less understood are nonpathogenic microorganisms that inhabit the phyllosphere. The composition of the phyllosphere microbiota has been analyzed in only a few studies by cultivation-independent methods; however, such methods are essential in light of the yet uncultivated majority of bacteria existing in nature, or more specifically on plant leaves. Not only their identity, but in particular the physiological properties of phyllosphere bacteria, their adaptations to the habitat, and their potential role (*e.g.*, with respect to modulating population sizes of pathogens) remain largely unknown. Current knowledge on the traits important in the phyllosphere is derived from relatively few studies on gene expression and stems mostly from model bacteria cultivated on host plants under controlled conditions.

However, under natural conditions, plants and their residing microorganisms are exposed to a host of diverse, highly variable environmental factors, including UV light, temperature, and water availability; moreover, individual microbes are subjected to competition with other microorganisms over resources, such as nutrients and space.

Toward a deeper understanding of phyllosphere microbiology, and in particular to learn more about the commensal majority of plant leaf colonizing bacteria, which may be of relevance for plant health and development, integrated approaches are needed. Here, Bacterial communities in the phyllosphere are thought to be limited by carbon availability, and it may be expected that access to carbon compounds on leaves is a major determinant of epiphytic colonization. There is evidence that small amounts of nutrients, such as simple sugars including glucose, fructose, and sucrose, leach from the interior of the plant.

The above-ground parts of plants are normally colonized by a variety of bacteria, yeasts, and fungi. While a few microbial species can be isolated from within plant tissues, many more are recovered from the surfaces of healthy plants. The aerial habitat colonized by these microbes is termed the phyllosphere, and the inhabitants are called epiphytes. While there has been some investigation of the colonists of buds and flowers, most work on phyllosphere microbiology has focused on leaves, a more dominant aerial plant structure. Bacteria are by far the most numerous colonists of leaves, often being found in numbers averaging 10^6 to 10^7 cells/cm2 (up to 10^8 cells/g) of leaf.

Compared to most other bacterial habitats, there has been relatively little examination of phyllosphere microbiology. This is somewhat surprising given the abundance of plants in the world and the roles of various phyllosphere bacteria in the important processes discussed below. Leaves constitute a very large microbial habitat.

It is estimated that the terrestrial leaf surface area that might be colonized by microbes is about 6.4×10^8 km^2. Given the large number of bacteria on leaves in temperate regions of the world and that population in tropical regions are probably even larger, the planetary phyllosphere bacterial population may be as large as 10^{26} cells. Clearly, in aggregate, these bacteria are sufficiently numerous to contribute in many processes of importance to global processes, as well as to the behavior of the individual plants on which they live. The microbial communities of leaves are diverse and include many different genera of bacteria, filamentous fungi, yeasts, algae, and, less frequently, protozoa and nematodes. Filamentous fungi are considered transient inhabitants of leaf surfaces, being present predominantly as spores, whereas rapidly sporulating species and yeasts colonize this habitat more actively.

Bacteria are by far the most abundant inhabitants of the phyllosphere. Epiphytic bacterial populations differ sharply in size among and within plants of the same species, as well as in close proximity, and over short time scales as well as over the growing season. These considerable variations in population sizes are caused in great part by the large fluctuations in the physical and nutritional conditions characteristic of the phyllosphere. Additionally, plant species appear to influence the microbial carrying capacity of the leaf, since the total number of culturable bacteria recovered from broad-leaf plants such as cucumber and beans was significantly greater than that recovered from grasses or waxy broad-leaf plants. Reflective of marked differences in the physicochemical environments of above-ground versus subterranean plant surfaces, the leaf bacterial flora differs substantially from that of roots. For example, pigmented bacteria, which are rarely found in the rhizosphere, dominate leaf surfaces, presumably because solar radiation influences the ecology of the phyllosphere. The differential composition of leaf and root bacterial communities is further evidenced by the failure of common root colonizers such as Rhizobium and Azospirillum to become established on leaves.

Studies of the composition of bacterial communities on leaves have been numerous but rather limited in scope. It is generally believed that populations of culturable aerobic bacteria on leaves are dominated by a few genera. A few exhaustive studies of the variations in the microbial community of leaves over multiple time and space scales have provided important detailed knowledge about the identity and the ecology of bacterial leaf inhabitants. Ercolani made an extensive inventory of culturable aerobic bacteria isolated from the surface of olive leaves over six growing seasons and reported distinct bacterial community structures on leaves of the same age at a given time of the growing season.

Most importantly, like Ercolani, they found distinct patterns of microbial colonization at different times of the year, with bacterial community diversity being lowest during the warmest and driest months of the season and highest during the cooler and rainy months. Coincidentally, in both of the above described studies, communities on young leaves were composed of a greater number of taxa than those of old leaves. Thus, specific natural environments of the phyllosphere apparently select for the presence of specific genotypes within the leaf bacterial community. This is further supported by the finding that the acquisition by Pseudomonas fluorescens of plasmids that are indigenous to the leaf microflora coincided with a specific maturation stage of the plant over two consecutive years. This indicated that traits carried on these plasmids conferred variable selective fitness to specific plasmid-bacterial host combinations during the growing season, possibly in response to changing conditions in the phyllosphere habitat.

The study of bacterial colonizers of leaves has been restricted mostly to aerobic culturable bacteria and also driven by the importance of investigating the ecology of plant-pathogenic bacteria because of their deleterious effect on plant productivity. Thus, the microbial ecology of the phyllosphere has been viewed mainly through the biology of gram-negative bacteria such as Pseudomonas syringae and Erwinia (Pantoea) spp., two of the most ubiquitous bacterial participants of phyllosphere communities. There is reason to believe, however, that the extreme fluctuations in the physicochemical environment of the phyllosphere over short time scales may select for bacterial species that have unusual and versatile traits that make them fit to colonize plant surfaces but have remained unculturable. The leaf surface has long been considered a hostile environment for bacterial colonists. The leaf surface is exposed to rapidly fluctuating temperature and relative humidity, as well as repeated alternation between presence and absence of free moisture due to rain and dew. The leaf also provides limited nutrient resources to bacterial colonists. While other habitats probably offer more extreme conditions of desiccation or temperature, *etc.*, they may not be subject to such rapid and extreme fluctuations in these several physical conditions. Several factors may influence the microhabitat experienced by bacteria on leaves. First, the leaf itself is surrounded by a very thin laminar layer in which moisture emitted through stomata may be sequestered, thereby alleviating the water stress to which epiphytes are exposed.

Second, some cells in a leaf bacterial population, particularly in plant-pathogenic populations, may not reside in exposed sites on the leaf surface but instead may at least locally invade the interior of the leaf, avoiding the stresses on the exterior of the leaf by residing in substomatal chambers or other interior locations. Thus, while some phytopathogens may have the option of avoiding stresses, most other epiphytes apparently must tolerate them in some way.

The phyllosphere has many features that make it an excellent habitat in which to study microbial ecology. Leaves are clean, and microbes can be observed directly on leaves, enabling the use of powerful new microscopic techniques to measure microbial identity, activity, and gene expression. Plants can be readily grown without epiphytic microbial communities, allowing us to readily manipulate their inhabitants, while communities can be made as simple or complex as needed by simple inoculation. In addition, important microbial processes, such as immigration, and ecological models, such as island biogeography, can be readily explored in epiphytic bacterial systems. Thus, phyllosphere microbiology has much to offer to the field of microbial ecology and promises to contribute to more effective and less environmentally damaging means of plant protection.

Methylotrophs

Methylotrophs are a diverse group of microorganisms that can use reduced one-carbon compounds, such as methanol or methane, as the carbon source for their growth; and multi-carbon compounds that contain no carbon bonds, such as dimethyl ether and dimethylamine. This group of microorganisms also includes those capable of assimilating reduced one-carbon compounds by way of carbon dioxide using the ribulose bisphosphate pathway.

These organisms should not be confused with methanogens which on the contrary produce methane as a byproduct from various one-carbon compounds such as carbon dioxide. Some methylotrophs can degrade the greenhouse gas methane, and in this case they are called methanotrophs. The methanotroph Methylococcuscapsulatus is used to degrade methane and other pollutants. The abundance, purity, and low price of methanol compared to commonly used sugars make methylotrophs competent organisms for production of amino acids, vitamins, recombinant proteins, single-cell proteins, co-enzymes and cytochromes.

Methylotrophic organisms are able to grow at the expense of one or more of the compounds typically used by methophiles, but cannot use methane. Methylotrophs are a diverse group, including both Gram-negative and Grampositive genera. None of them make resting structures like exospores and cyst and none of them have the complex intracellular membrane systems that characterize methanotrophs growing on methane

There are two sub groups:

1- Obligate methylotrophs.

2- Facultative methylotrophs.

Obligate methylotrophs

A single obligate methylotroph (methylophilus) is known. It is Gram-negative, polarly flagellated rod capable of rapid growth with methanol. Some strains can also utilize formaldehyde or methylamines. Carbon is assimilated via the ribulose mono phosphate pathway.

Facultative methylotrophs

It is relatively widely distributed trait among heterotrophic bacteria. It may also be common among chemoautotrophs: several thiobacilli and nitrifying bacteria can drive CO_2 assimilation via the Calvin-Benson cycle by formate oxidation.

Examples of methylotrophs:

Genus Methanosarcina

Methylococcus capsulatus

Pichia pastoris

Bacteria of the genus Methylobacterium are well-studied examples of facultative methylotrophs. These bacteria are classified as proteobacteria and are capable of growth on methanol and methylamine as well as on a variety of C_2, C_3, and C_4 compounds. Bacteria often live in environments containing diverse substrates. One of such habitat is the phyllosphere, where facultative methylotrophic bacteria are found to be ubiquitous and abundant. These methylotrophic bacteria belong to the genus Methylobacterium and are known to metabolize methanol but also a limited number of alternative carbon substrates, such as organic acids and alcohols. Because of their distinctive pink pigmentation, they are sometimes referred to as PPFMs (pink-pigmented facultative methylotrophs).

PPFMs are aerobic, Gram-negative bacteria and, although they are able to grow on a wide range of multicarbon substrates, they are characterized by the capability to grow on one-carbon compounds such as formate, formaldehyde, and methanol as their sole carbon and energy source. They are easily isolated on a methanol based mineral medium. Recently published data suggest that the degree of the plant–Methylobacterium association varies from strong, or symbiotic to loose, or epiphytic, a range that also includes the intermediate endophytic association. The use of leaf impressions on a methanol-agar medium has demonstrated the presence of pink-pigmented facultative methylotrophs in over 70 plant species. These

PPFMs are especially abundant on leaves of field-grown crops averagedabout10^6 cfu of PPFMs per leaflet, and typically >80% of the viable bacteria recovered from leaves were PPFMs. The occurrence of soil methylotrophs is probably related to the abundance of plant lignin and pectin in soils; these polymers are major potential sources of methanol.

Plant leaf surfaces release diverse carbon sources, mainly sugars and organic acids, at low amounts (μM range), and these sources are heterogeneously located and result of leaching through the cuticle. In addition to these substrates, volatile carbon substrates, particularly the plant cell wall metabolism byproduct methanol, are released via the stomata. Methanol emission peaks in the morning, when the stomata first open. There is evidence that methanol is consumed by Methylobacterium and contributes to the epiphytic fitness of the organism. However, in addition to the peak of methanol emission in the morning, Methylobacterium should adapt its metabolism to use additional carbon sources during the rest of the day when methanol emission is low or during the night when stomata are closed and methanol is consequently no longer available.

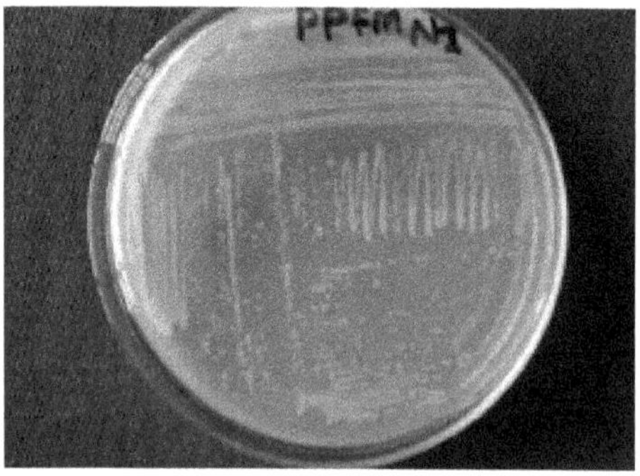

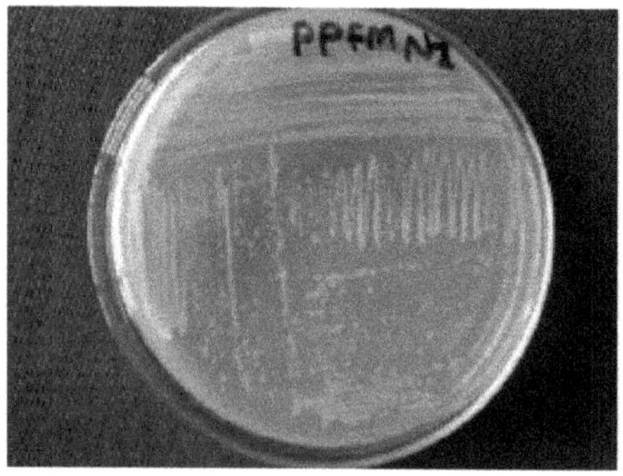

A number of reports suggest that Methylobacterium strains are more than passive passengers on plant leaves. For instance, they have been shown in some cases to stimulate seed germination and plant development, possibly by production of phytohormones and it has been reported that one strain produces the cytokinin zeatin while others have been reported to produce indole acetic acid. In addition, Methylobacterium strains have been suggested to contribute to the flavor of strawberries and have been localized as endosymbionts within cells of the buds of Scotch pine (Pinus sylvestris). One nonpigmented Methylobacterium strain has been shown to form a root-nodulating nitrogen-fixing symbiosis with a legume. They show that four different Methylobacterium strains representing leaf isolates and a type M. extorquens strain all produce the cytokinin trans-zeatin at low levels in pure culture and excrete it into the culture medium.

7

Plant – Microbe Interactions

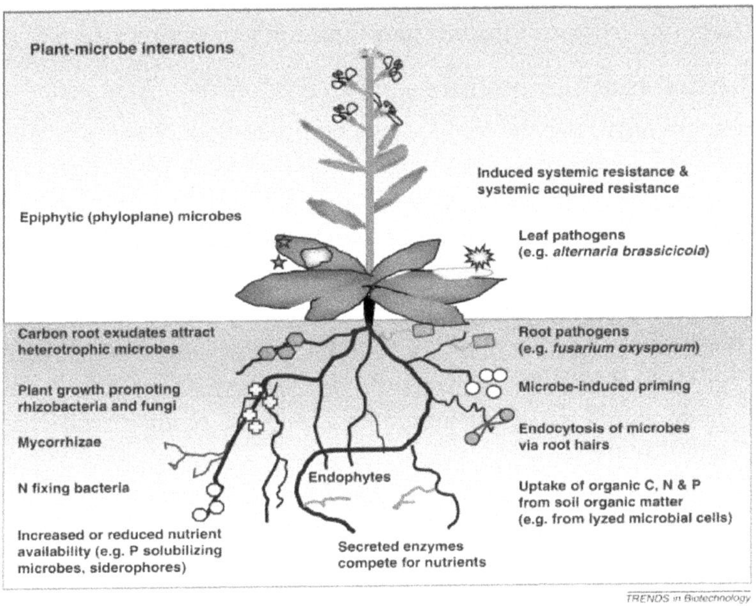

Symbioses

- Many microbes (bacteria, fungi) have important symbioses with plants
- Rhizosphere = thin layer of soil immediately attached to root hairs of plants. Typically contains 10^9 microbes/g of soil.
- Many rhizosphere organisms are ectosymbionts, living outside the roots. Others are endosymbionts, living inside or penetrating into plant roots.

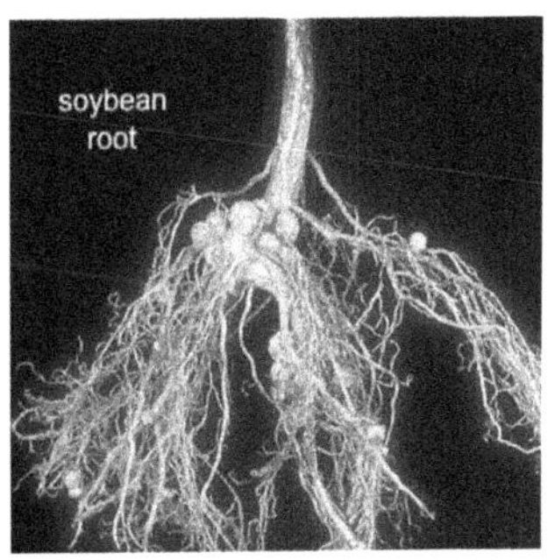

- Many of these bacteria contribute Nitrogen fixation, obtain plant nutrients in return (see below for Rhizobium symbiosis).

Rhizobium-Legume symbioses

- Plants of the legume family (soybeans, clover, alfalfa, beans, peas) can grow in soils lacking nitrogen compounds required by other plants.
- These plants contain endosymbiotic Rhizobium bacteria that grow in root nodules. Rhizobia can fix atmospheric Nitrogen gas (N_2)___$N_2 + 6[H]\ 2NH_3$
- The reaction requires total lack of oxygen and lots of energy as ATP. To bind oxygen and get rid of it, bacteria use protein called leghemoglobin, somewhat similar to animal hemoglobin. Globin part is encoded in plant genome, heme group is encoded in bacterial genome.
- Neither partner can fix nitrogen alone, only in symbiosis.

Hydrothermal vent Communities

- Occurs only near thermal springs on ocean floor, 2 miles or more below surface. Totally black, no sunlight penetrates below 600 feet.
- Associated with spreading centers of tectonic plates where hot magma close to surface causes area of floor to slowly drift apart.
- Seawater seeps down, mixes w/ minerals at high temperature comes back to ocean water in plumes at 270-380 deg. C. These are sometimes called black smokers since minerals precipitate as black cloud when in contact with cold sea water.
- Contains high levels of inorganics: Mn_{2+}, H_2, usually H_2S; very low in organic matter
- Astonishing discovery: such regions are densely populated by a community of unusual animals: 2 m long tube worms, giant clams, mussels, white shrimp.
- What do they eat? Unlike earth's surface, there is no source of light to stimulate phototrophs, they "eat" chemolithotrophic bacteria!
- Example: Inside the tube worms live huge colonies of bacterial endosymbionts. These are autotrophic chemolithotrophs, oxidizing sulfide to sulfate as their energy source. As bacteria grow, they provide carbon and nitrogen compounds for worms to feed on.

Ruminant Symbiosis

- Ruminants are the herbivorous mammals whose digestive tract contains four chambers. First chamber (largest) is the rumen, provides a place for bacteria to break down the fiber in the plants so the cow can use it for energy.
- Includes cows, sheep, giraffes, buffalo, and *etc.*
- Ruminants eat grasses and other plant materials, but do not produce enzymes todigest cellulose, the primary plant metabolite.

- Instead, ruminants rely on huge microbial community in rumen to digest plant materials. Microbial densities can reach as high as 10^{12} microorganisms/ml, the highest density found anywhere in nature.
- Ruminants feed off fermentation waste products of microorganisms; mainly acetic acid, propionic acid, and butyric acid.

Gnotobiotic Animals

- Gnotobiotic = "known microbiota"; animal host is either entirely free of microbes (aka "germfree", "axenic") or has a microbiota whose identity is completely known.
- Animals in utero are germfree, but acquire resident bacteria within hours of birth.
- Relatively easy to produce germfree animals for birds. Sterilize shell, use sterile incubator, keep animals in an environment where all air, food, water is sterilized before entry.
- More difficult to establish germfree animals other than birds. Need cesarean section of pregnant females, germfree isolation chambers where all air, food, water is sterilized before entry.
- Germfree animals generally are less healthy than animals with normal microbiota. Defects include: Greater vitamin requirements for K and B complex, lower cardiac output much more susceptible to pathogens -- normal microbiota colonize access sites, often compete successfully to prevent pathogens from binding to host tissues much smaller infectious dose required to initiate an infection

Interrelationship between microorganisms: Beneficial and harmful relationships

Interrelationships in soils are of 3 types

1. Plant microbe interaction
2. Microbe – microbe interaction and
3. Plant microbe – microbe interaction

a) Plant microbe interaction

It mainly constitutes the association of microorganism with plants little in a positive way or in a negative way. The positive approach is mainly the symbiotic relationships and the negative approach constituents mainly pathogen plant interactions.

b) Plant microbe - microbe interaction

Also called tripartite symbiosis

Eg: Alnus – Frankia –Mycorrhiza and Casbarina – Frankia – Mycorrhiza

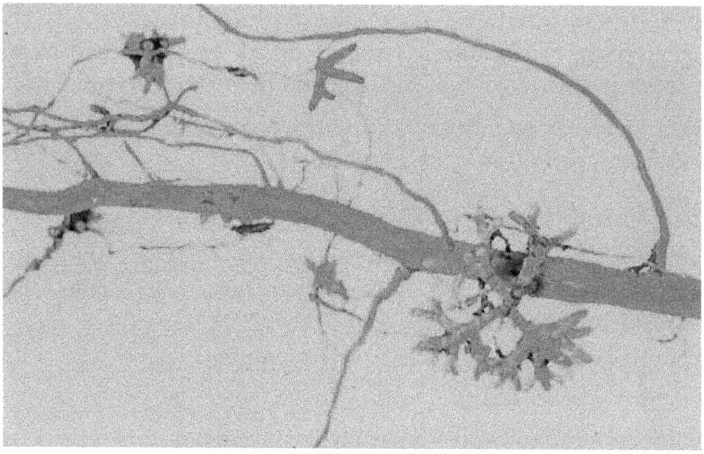

***Ceanothus* roots, with *Frankia* vesicles**

c) Microbe – microbe interaction

Microbial interaction in soil

Ecological Association/Interactions among Soil Microorganisms

Soil is the largest terrestrial ecosystem where a wide variety of relationships exists between different types of soil organisms. The associations existing between different soil microorganisms, whether of a symbiotic or antagonistic nature, influence the activities of microorganisms in the soil. Microflora composition of any habitat is governed by the biological equilibrium created by the associations and interactions of all individuals found in the community. In soil, many microorganisms live in close proximity and interact among them-selves in a different ways. Some of the interactions or associations are mutually beneficial, or mutually detrimental or neutral. The various types of possible interactions/associations occurring among the microorganisms in soil can be: a) beneficial i) mutualism ii) commensalisms and iii) proto-cooperation or b) detrimental / harmful - i) amensalism, ii) antagonism, iii) competition iv) Parasitism and v) predation

a) Beneficial Association/Interactions

1. Mutualism (Symbiosis): It is a relationship or a type of symbiosis in which both the interacting organisms/partners are benefited from each other. The way/ manner in which benefit is derived depends on the type of interactions. When the benefit is in the term of exchange of nutrients, then the relationship is termed as **"syntrophism"** (Greek meaning: Syn -mutual and trophe = nourishment), for example, Lichen (association of algae or BGA with fungus) in which algae benefits by protection afforded to it by the fungal hyphae from environmental stresses, while the fungus obtain and use CO_2 released by the algae during photosynthesis. Where the blue green algae are the partners in the lichen association, the heterotrophs (Fungus), benefit from the fixed nitrogen by the blue green algae.

Microorganisms may also form mutualistic relationships with plants, for example nitrogen fixing bacteria *i.e.* Rhizobium growing in the roots of legumes. In this Rhizobium-legume association, Rhizobium bacteria are benefited by protection from the environmental stresses while in turn plant is benefited by getting readily available nitrate nitrogen released by the bacterial partner.

The Anabaena-Azolla is an association between the water fern Azolla and the cyanobacterium Amabaena. This association is of great importance in paddy fields, where nitrogen is frequently a limiting nutrient.

An actinorrhizal symbiosis of actinomycetes, Frankia with the roots of Alnus and Casurina (non-legumes) is common in temperate forest ecosystem for soil nitrogen economy. Another type of symbiotic association which exists between the roots of higher plants and fungus is Mycorrhiza. In this association fungus gets essential organic nutrients and protection form roots of the plants and allows them to multiply and in turn plants uptake phosphorus, nitrogen and other inorganic nutrients made available by the fungus.

2. Commensalisms: In this association one organism/partner in association is benefited by other partner without affecting it. For example, many fungi can degrade cellulose to glucose, which is utilized by many bacteria. Lignin which is major constituent of woody plants and is usually resistant to degradation by most of the microorganisms but in forest soils, lignin is readily degraded by a group of Basidiomycetous fungi and the degraded products are used by several other fungi and bacteria which can not utilize lignin directly. This type of association is also found in organic matter decomposition process.

3. Proto-cooperation: It is mutually beneficial association between two species / partners. Unlike symbiosis, proto-cooperation is not obligatory for their existence or performance of a particular activity. In this type of association one organism favor its associate by removing toxic substances from the habitat and simultaneously obtain carbon products made by the another associate/partner. Nutritional proto-cooperation between bacteria and fungi has been reported for various vitamins, amino and purines in terrestrial ecosystem and are very useful in agriculture. Proto-cooperative associations found beneficial in agriculture are : i) synergism between VAM fungus-legume plants and Rhizobium in which nitrogen fixation and phosphorus availability / uptake is much higher resulting in higher crop yields and improved soil fertility, ii) synergism between PSM-legume plants and Rhizobium and iii) synergism between plant roots and PGPR in rhizosphere where rhizobacteria restrict the growth of phytopathogens on plant roots and secrets growth promoting substances.

b) Detrimental (Harmful) Associations/Interactions

1. Antagonism: It is the relationship in which one species of an organism is inhibited or adversely affected by another species in the same environment. In such antagonism, one organism may directly or indirectly inhibit the activities of the other. Antagonistic relations are most common in nature and are also important for the production of antibiotics. The phenomenon of antagonism may be categorized into three *i.e.* antibiosis, competition and exploitation.

In the process of antibiosis, the antibiotics or metabolites produced by one organism inhibits another organism. An antibiotic is a microbial inhibitor of biological origin. Innumerable examples of antibiosis are found in soil. For example, Bacillus Species from soil produces an antifungal agent which inhibits growth of several soil fungi. Several species of Streptomyces from soil produces antibacterial and antifungal antibiotics. Most of the commercial antibiotics such as streptomycin, chloramphenicol, Terramycin and cyclohexamide have been produced from the mass culture of Streptomyces. Thus, species of Streptomyces are the largest group of antibiotic producer's in soil. Another example of antibiosis is inhibition of Verticillium by Trichoderma, inhibition of Rhizoctonia by a bacterium Bacillus subtilis, inhibition of soil fungus Aspergillus terreus by a bacterium Staphylococcus aureus.

2. Ammensalism: In this interaction /association one partner suppress the growth of other partner by producing toxins like antibiotics and harmful gases like ethylene, HCN, Nitrite *etc.*

3. Competition: As soil, is inhabited by many different species of microorganisms, there exists an active competition among them for available nutrients and space. The limiting substrate may result in favoring one species over another. Thus, competition can be defined as "the injurious effect of one organism on another because of the removal of some resource of the environment". This phenomenon can result in major fluctuations in the composition of the microbial population in the soil.

For example, chlamydospores of Fusarium, Oospores of Aphanomyces and conidia of Verticillium dahlae require exogenous nutrients to germinate in soil. But other fungi and soil bacteria deplete these critical nutrients required for spore germination and thereby hinder the spore germination resulting into the decrease in population. Competition for free space has been, reported to suppress the fungal population by soil bacteria. Therefore, organisms with inherent ability to grow fast are better competitors.

4. Parasitism: It is an association, in which one organism lives in or on the body of another. The parasite is dependent upon the host and lives in intimate physical contact and forms metabolic association with the host. So this is a host -parasite relationship in which one (parasite) is benefited while other (host) is adversely affected, although not necessarily killed. Parasitism is widely spread in soil communities, for example, bacteriophages (viruses which attack bacteria) are strict intracellular parasites Chytrid fungi, which parasitize algae, as well as other fungi and plants; there are many strains of fungi which are parasitic on algae, plants, animals parasitized by different organisms, earthworms are parasitized by fungi, bacteria, viruses *etc.*

5. Predation: Predation is an association / exploitation in which predator organism directly feed on and kills the pray organism. It is one of the most dramatic inter relationship among microorganisms in nature, for example, the nematophagous fungi are the best examples of predatory soil fungi. Species of Arthrobotrytis and Dactylella are known as nematode trapping fungi. Other examples of microbial predators are the protozoa and slime mold fungi which feed on the bacteria and reduce their population. The bacteriophages may also be considered as predators of bacteria.

8

Bio-Geochemical Cycling

All organisms require the six elements **CHNOPS** in sizable quantities, and that these minerals may occur in different forms. In fact, depending on whether the local environment is aerobic or anaerobic, the exact chemical form of any of these elements may change quickly and drastically. Example: If a soil becomes water logged and anaerobic, sulfate will be converted into hydrogen sulfide and nitrate into nitrogen gas or ammonia (via anaerobic respiration). As water recedes and the soil becomes aerobic again, the ammonia is oxidized back to nitrate in stages. All of this (and many other processes) occur largely because of the activities of soil microbes. Without these microbes, life as we know it would not occur, because necessary elements would remain tied up in unusable chemical forms and gradually be removed by sedimentation. To maintain the level of these components it is essential that they undergo a regular process of recycling. Different microorganisms bring about this process of recycling through various transformations.

Soil Microorganisms in Cycling of Elements or Plant Nutrient

Soil microorganisms are the most important agents in the cycling / transformation of various elements (N, P, K, S, Iron *etc.*) in the biosphere; where the essential elements undergo cyclic alterations between the inorganic state as free elements in nature and the combined state in living organisms. Life on earth is dependent on the cycling of nutrient elements from their elemental states to inorganic compounds to organic compounds and back into their elemental states.

The microbes through the process of biochemical reactions convert / breakdown complex organic compounds into simple inorganic compounds and finally into their constituent elements. This process is known as **"Mineralization"**.

Mineralization of organic carbon, nitrogen, phosphorus, sulphur and iron by soil microorganisms makes these elements available for reuse by plants. In the following paragraphs the cycling / transformations of some of the important elements are discussed.

The four most important cycles are mention below

- Nitrogen Cycle
- Sulphur Cycle / Sulphur Transformation

- ❑ Phosphorus Cycle / Transformation
- ❑ Iron Cycle / Transformation

Nitrogen Cycle

Although molecular nitrogen (N_2) is abundant (i.e 78-80 % by volume) in the earth's atmosphere, but it is chemically inert and therefore, cannot be utilized by most living organisms and plants. Plants, animals and most microorganisms, depend - on a source of combined or fixed nitrogen (*e.g.* ammonia, nitrate) or organic nitrogen compounds for their nutrition and growth. Plants require fixed nitrogen (ammonia, nitrate) provided by microorganisms, but about 95 to 98 % soil nitrogen is in organic form (unavailable) which restrict the development of living organisms including plants and microorganisms. Therefore, cycling/transformation of nitrogen and nitrogenous compounds mediated by soil microorganisms is of paramount importance in supplying required forms of nitrogen to the plants and various nutritional classes of organisms in the biosphere. In nature, nitrogen exists in three different forms *viz.* gaseous / gas (78 to 80 % in atmosphere), organic (proteins and amino acids, chitins, nucleic acids and amino sugars) and inorganic (ammonia and nitrates).

Biological N_2 Fixation

A. Symbiotic: *e.g.* Rhizobium (Eubacteria) legumes, Frankia (Actinomycete) and Anabaena (cyanobacteria) non - legumes

B. Non Symbiotic

1. **Free Living:** *e.g.* Azotobacter, Derxia, Bejerinkia, Rhodospirillum and BGA.
2. **Associative:** *e.g.* Azospirillum, Acetobacter, Herbaspirillim.

Nutritional categories of N_2 fixing Bacteria

1. **Heterotrops**
2. **Photoautotrophs**

Nitrogen cycle is the sequence of biochemical changes form free atmospheric N_2 to complex organic compounds in plant and animal tissues and further to simple inorganic compounds (ammonia, nitrate) and eventual release of molecular nitrogen (N_2) back to the atmosphere is called "nitrogen cycle".

In this cycle a part of atmospheric nitrogen (N_2) is converted into ammonia and then to amino acids (by soil microorganisms and plant-microbe associations) which are used for the biosynthesis of complex nitrogen-containing organic compound such as proteins, nucleic acids, amino sugars *etc.* The proteins are then degraded to simpler organic compounds *viz.* peptones and peptides into amino acids which are further degraded to inorganic nitrogen compounds like ammonia, nitrites and nitrates. The nitrate form of nitrogen is mostly used by plants or may be lost through leaching or reduced to gaseous nitrogen and subsequently goes into the atmosphere, thus completing the nitrogen cycle. Thus, the process of mineralization (conversion of organic form of nutrients to its mineral /inorganic form) and immobilization (process of conversion of mineral / inorganic form of nutrient elements into organic form) are continuously and simultaneously going on in the soil.

Biological availability of N, P and K is of considerable economic importance, since they are the major plant nutrients derived from the soil. Of the three, N stands out as the most susceptible one to microbial transformations. This element is the key building block of the protein molecule upon which all life is based; it is an indispensable component of the protoplasm of plants, animals and microorganisms.

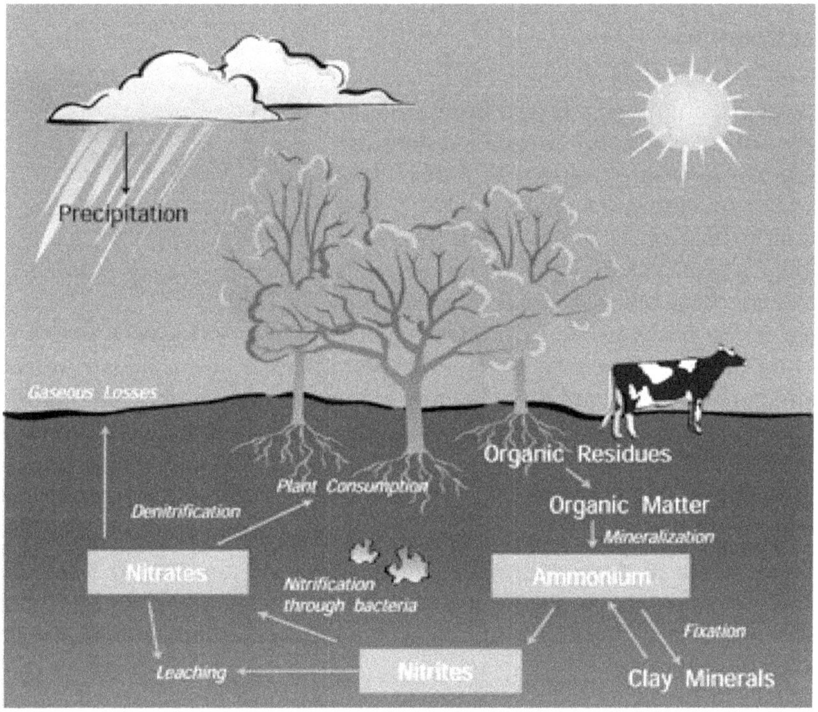

Molecular N_2 constitutes about 78% of the earth's atmosphere but it is chemically inert and cannot be utilized by more living organisms, plants, animals and microorganisms. Therefore they depend on a source of combined N such as ammonia, nitrate or organic N compounds for their growth.

Nitrogen undergoes a number of transformations that involve organic, inorganic and volatile forms of nitrogen. A small part of the large reservoir of N_2 in the atmosphere is converted to organic compounds by certain free living microorganisms or by plant microbe association that makes the element available to plant growth. The nitrogen present in the proteins or nucleic acids of plant tissue is used by animals. In the animal body, the N is converted to other simple and complex compounds. Upon the death, plants and animals undergo microbial decay and organic N is released as ammonium, which is then utilized by vegetation or is oxidized to nitrate by microorganisms. The nitrate form of N is mostly used by the plants or may be lost by bacteria reduced to gaseous N_2, which escapes to atmosphere, there by completing the cycle.

The Nitrogen cycle mainly includes the following transformations such as

- Mineralization
- Nitrification
- Denitrification
- Immobilization
- Nitrogen Fixation

Mineralization

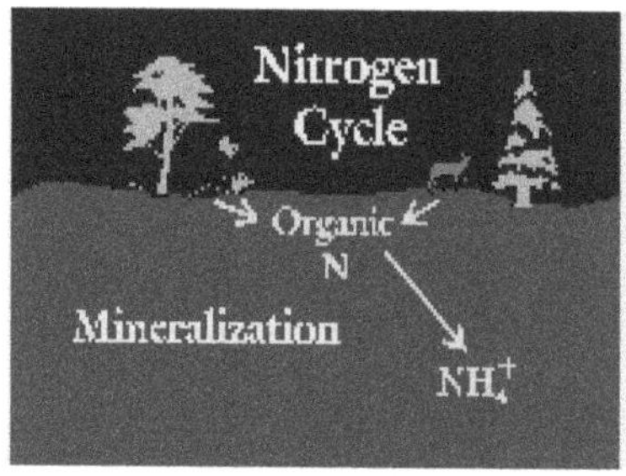

It is the process in which proteins, nucleic acids, amino sugars and other organic compounds of plant and animal residues are degraded by microorganisms with the eventual liberation of ammonia. This is also called as ammonification. A part of the liberated ammonia is assimilated by the microorganisms themselves. The first step of ammonification process is the hydrolysis of proteins, nucleic acids and other organic nitrogenous compounds into amino acids (Proteolysis). The amino compounds are then deaminated to yield ammonia. Ammonification usually occurs under aerobic conditions while under anaerobic conditions protein decomposition leads to conversion of ammonia into amines and related compounds ex. Clostridium. The anaerobic decomposition of protein is called as putrefaction. These amines are subsequently oxidized in the presence of O_2 to release ammonia.

Break down of nitrogenous substance is brought about by the activity of a multitude of microbial species. Almost all bacteria, actinomycetes and fungi can bring about proteolysis and the amino acids produced are utilized for the growth of these organisms.

Nitrification

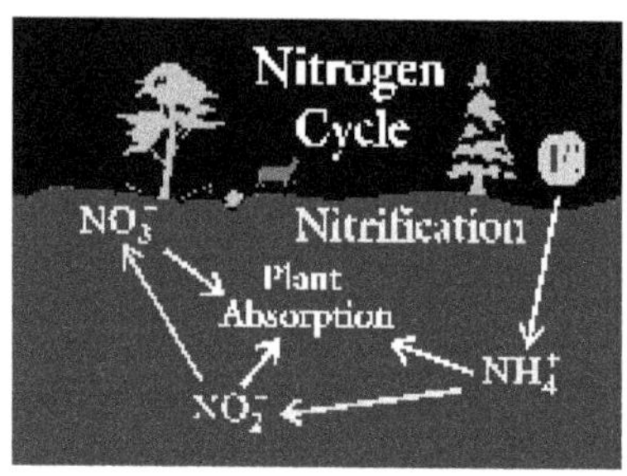

The biological oxidation of ammonium salts (in soil) to nitrites and the subsequent oxidation of nitrites to nitrates is called as nitrification. *i.e.* the biological convention of N in soil from a reduced to a more oxidized state, called nitrification. This process occurs primarily by a group of aerobic chemoautotrophs called the nitrifiers. While some low levels of heterotrophic nitrification do occur, rates are low and quantities of nitrate produced are relatively small compared to the chemoautotrophs. All nitrifiers are aerobic, and nitrification proceeds at C: N ratios of less than 20, when N is abundant. In some settings, such as forest litter layers, most nitrification occurs by saprophytic fungi rather than chemoautotrophic bacteria. Nitrification is a two-step process.

First Step: **Ammonium Oxidation:**

Ammonia is oxidized to nitrite.

$2NH_3 + 1\frac{1}{2}H_2O_2 \rightarrow NO_2^- + 2H + H_2O$ - Nitrosofication

The microorganisms involved are called ammonia oxidizers. This change is brought about by chemoautotrophic bacteria of the genera Nitrosomonas, Nitrosolobus, Nitrosococcus, Nitrosospira. These bacteria obtain their energy requirement by the oxidation of NH_4^+ to NO^-_2. Among the nitrifiers, Nitrosomonas are most important in soils. Nitrosospira is an aquatic nitrifier.

Second step: **Nitrite Oxidation**

Nitrite is further oxidized to nitrate.

$HNO_2 + \frac{1}{2}O_2 \rightarrow HNO_3$.

The Organisms involved are Nitrobacter, Aspergillus, Penicillium, Cephalosporium.

Denitrification

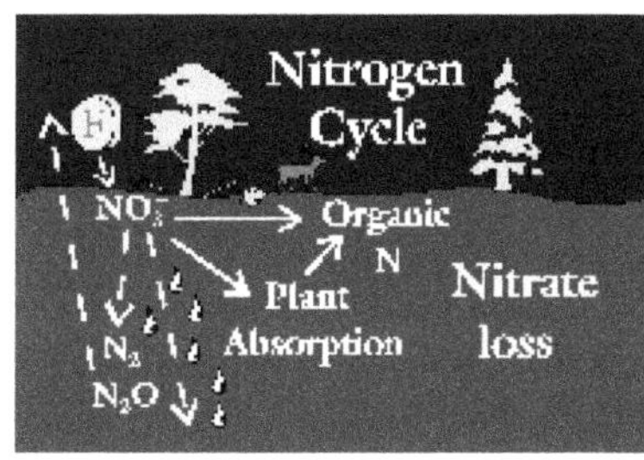

The conversion of nitrate and nitrite into molecular N_2 or nitrous oxide through microbial processes is known as denitrification. Certain bacteria are capable of using nitrate as the terminal electron acceptor under anaerobic conditions. This is called **nitrate respiration**. As a consequence of nitrate respiration, NO_3 is reduced to N_2 gas or nitrous oxide. Denitrification leads to the loss of N from the soil. It depletes N, and therefore it is not a desirable reaction. The escape of molecular N into the atmosphere is also known as **volatalization**.

Denitrification occurs mostly in waterlogged anaerobic soils with a high organic matter contents. Denitrification of bound nitrogen to gaseous N is mediated by numerous species of bacteria, which normally use O_2 as hydrogen acceptor (aerobically) and, also use nitrates and nitrites (anaerobically). Anaerobic conversion of nitrate into molecular nitrogen is known as nitrate respiration. Bacterial genera that bring about denitrification are Pseudomonas, Achromobacter, Bacillus, Micrococcus.

$2NO^-_3 + 10\,H \rightarrow N_2 + 4H_2O + 2OH^-$ (or)

$2NO^-_2 + 6\,H \rightarrow N_2 + 2H_2O + 2OH^-$ (or)

$N_2O + 2H \rightarrow N_2 + H_2O$

Since nitrates are used as a source of electron acceptor, there is a net loss of N from soil. This process is termed also as **dissimilatory nitrate reduction**. Many soil bacteria like Thiobacillus denitrificans oxidize S (chemoautotrophically) and also reduce nitrate to nitrogen.

$5 S + 6 KNO_3 + 2 H_2O \rightarrow 3N_2 + K_2SO_4 + 4KHSO_4$ (or)

$5 K_2S_2O_3 + 8 KNO_3 + H_2O \rightarrow 4N_2 + 9 K_2SO_4 + H_2SO_4$

The reverse of nitrification process is termed as **nitrate reduction**. That is the reduction of nitrate to nitrite and then ammonia.

$HNO_3 + 4H_2 \rightarrow NH_3 + 3H_2O$

Since organisms are able to obtain cellular N through ammonia assimilation, the process is called as **assimilatory nitrate reduction.**

General pathway of denitrification

Nitrate is first reduced to nitrite, which is then transformed to nitrous oxide (NO). The nitrous oxide is converted to N_2 with N_2O as an intermediate.

$$2\ HNO_3 \xrightarrow{1} 2HNO_2 \xrightarrow{2} 2\ NO \xrightarrow{3} N_2O \xrightarrow{4} N_2$$

The enzymes involved in the above reactions are

1. Nitrate reductase
2. Nitrite reductase
3. Nitric oxide reductase
4. Nitrous oxide reductase

Fallow soils flooded with water are more congenial for denitrification than well drained and continuously cropped soils. Though it is an undesirable reaction in the point of view of plant nutrition, it has ecological importance. Because, with out denitrification the supply of N on the earth would have got depleted and NO_3 would have accumulated. High concentration of NO_3 are toxic, denitrification is a mechanism by which some of the N is released back to the atmosphere. It is essential to distinguish nitrification, denitrification as separate processes, and the difference between these are furnished below in the table.

	Nitrification	Denitrification
Aerobic status	Obligate Aerobes	Facultative
Respiratory Classification	Aerobic Respiration	Anaerobic Respiration
Respiratory electron acceptor	O_2	NO_3^-, NO_2^-, others
C-source	Autotroph	Heterotroph
Energy source	Chemolithotroph	Chemoorganotroph
Taxonomic group	Nitrosomonas, Nitrobacter	Variable group; many different genera
Reversibility	Reversible	Not reversible

Immobilization

In nitrogen immobilization, ammonia and nitrate are taken up by microbes and is largely immobilized or made unavailable to plants, depending on the C: N ratios.

When N is abundant, both microbes and plants assimilate ammonia and nitrate. In contrast to mineralization, microbial immobilization leads to the biosynthesis of the complex molecules of microbial protoplasm from ammonium and nitrate. Immobilization results in a marked depression of nitrogen uptake by the plant. The mineralization of organic N and the microbial assimilation of inorganic ions proceed simultaneously in soil.

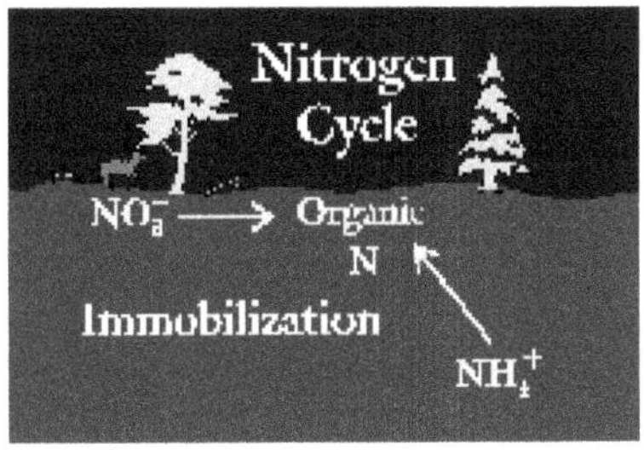

Relationship between the C: N ratios and net immobilization and nitrification are depicted in the table below:

< 20	20-30	> 30
Net gain of NH4+ and NO3-	Neither gain nor loss	Net uptake of NH4+ and NO3-

The C: N is used as an indicator to know the step in the nitrogen cycle. The ratios less than 20 means, excess N is present and nitrification proceeds (with a net gain of N). With ratios between 20 and 30, nitrification and immobilization rates are in equilibrium and there is no net gain or loss of N. With a ratio greater than 30, N is limited and net immobilization occurs with uptake (or loss) of N from the active N cycle.

Biological Nitrogen Fixation

Molecular nitrogen (N_2) is the major component (approximately 80%) of the earth's atmosphere. The element nitrogen is an essential part of many of the chemical compounds, such as proteins and nucleic acids, which are the basis of all life forms. However, N_2 cannot be used directly by biological systems to build the chemicals required for growth and reproduction. Before its incorporation into a living system, N_2 must first be combined with the element hydrogen. This process of reduction of $N_{2,}$ commonly referred to as "nitrogen fixation" (N-fixation).

Early in studies of nitrogen fixation, it was expected that biological extracts could produce ammonia if provided with nitrogen gas and a strong biological reductant:

$N_2 + 6e^- + 6H^+$ -----> $2NH_3$

However, without further supplements, the reaction does not occur. It was discovered that ATP hydrolysis was necessary to drive the reaction. In addition, hydrogen gas is produced. The biological mechanism of nitrogen fixation requires eight electrons rather than six electrons, the extra two being used to produce the hydrogen gas:

$N_2 + 8e^- + 8H^+ + ATP$ -----> $2NH_3 + H_2 + ADP + Pi$

Two points about the energetics of biological nitrogen fixation: (i) The mechanism is somewhat wasteful - reducing power is lost in the formation of hydrogen gas. Although some microbes recycle the evolved hydrogen, many others do not. Four more electrons are required to recycle the hydrogen gas, for the final reaction:

$N_2 + 12\,e^- + 12\,H^+ + 12\,ATP \text{--->} 2NH_4 + 12\,ADP$--->Amino Acids--->Proteins

Overall Reaction

Step 1 $N_2 + 2e^- + 4ATP \text{---->} HN = NH$ (diimide)

Step 2 $HN = NH + 2e^- + 4ATP \text{---->} H_2N - NH_2$ (imide)

Step 3 $H_2N - NH_2 + 2e^- \text{---->} 2NH_4^+$ (ammonia)

Several biochemical steps involved in the nitrogen cycle are:

1. Proteolysis
2. Ammonification
3. Nitrification
4. Nitrate reduction
5. Denitrification.

1. Proteolysis: Plants use the ammonia produced by symbiotic and non-symbiotic Nitrogen fixation to make their amino acids & eventually plant proteins. Animals eat the plants and convert plant proteins into animal proteins. Upon death, plant and animals undergo microbial decay in the soil and the nitrogen contained in their proteins is released. Thus, the process of enzymatic breakdown of proteins by the microorganisms with the help of proteolysis enzymes is known as "proteolysis".

The breakdown of proteins is completed in two stages. In first stage proteins are converted into peptides or polypeptides by enzyme "proteinases" and in the second stage polypeptides / peptides are further broken down into amino acids by the enzyme "peptidases".

Proteins -----------------------> Peptides -----------------------> Amino Acids
Proteinases **Peptidases**

The amino acids produced may be utilized by other microorganisms for the synthesis of cellular components, absorbed by the plants through mycorrhiza or may be de animated to yield ammonia.

The most active microorganisms responsible for elaborating the proteolytic enzymes (Proteinases and Peptidases) are *Pseudomonas, Bacillus, Proteus, Clostridium Histolyticum, Micrococcus, Alternaria, Penicillium* etc.

2. Ammonification (Ammo acid degradation): Amino acids released during proteolysis undergo deamination in which nitrogen containing amino (-NH2) group is removed. Thus, process of deamination which leads to the production of ammonia is termed as "ammonification". The process of ammonification is mediated by several soil microorganisms. Ammonification usually occurs under aerobic conditions (known as oxidative deamination) with the liberation of ammonia (NH3) or ammonium ions (NH4) which are either released to the atmosphere or utilized by plants (paddy) and microorganisms or still under favorable soil conditions oxidized to form nitrites and then to nitrates.

The proce sses of ammonification are commonly brought about by *Clostridium* sp, *Micrococcus* sp, *Proteus sp.* etc. and it is represented as follows

CH_3 $CHNH_2$ COOH + 1/2 O_2 -----------------> CH3COCOOH + NH_3

Alanine Deaminase Pyruvic acid ammonia

3. Nitrification: Ammonical nitrogen / ammonia released during ammonification are oxidized to nitrates and the process is called "nitrification". Soil conditions such as well aerated soils rich in calcium carbonate, a temperature below 30 ° C, neutral PH and less organic matter are favorable for nitrification in soil.

Nitrification is a two stage process and each stage is performed by a different group of bacteria as follows.

Stage I: Oxidation of ammonia of nitrite is brought about by ammonia oxidizing bacteria *viz. Nitrosomnonas europaea, Nitrosococcus nitrosus, Nitrosospira briensis, Nitrosovibrio* and *Nitrocystis* and the process is known as nitrosification. The reaction is presented as follows.

$2NH_3$ + 1/2O_2 --------------------> NO_2 + 2H + H_2O

Ammonia Nitrite

Stage II: In the second step nitrite is oxidized to nitrate by nitrite-oxidizing bacteria such as *Nitrobacter winogradsky .Nitrospira gracilis, Nirosococcus mobiiis* etc, and several fungi*(eg. Penicillium, Aspergillus) and* actinomycetes *(eg. Streptomyces, Nocardia).*

NO_2(-) + ½O_2 ----------------------> NO_3

Nitrite ions Nitrate ions

The nitrate thus, formed may be utilized by the microorganisms, assimilated by plants, reduced to nitrite and ammonia or nitrogen gas or lost through leaching depending on soil conditions. The nitrifying bacteria (ammonia oxidizer and nitrite oxidizer) are aerobic gram-negative and chemoautotrophic and are the common inhabitants of soil, sewage and aquatic environment.

4. Nitrate Reduction: Several heterotrophic bacteria *(E. coli, Azospirillum)* are capable of converting nitrates to nitrites and nitrites to ammonia. Thus, the process of nitrification is reversed completely which is known as nitrate reduction. Nitrate reduction normally occurs under anaerobic soil conditions (water logged soils) and the overall process is as follows:

HNO_3 + $4H_2$ --------------------> NH4 + $3H_2O$

Nitrate Reductase Ammonium

Nitrate reduction leading to production of ammonia is called "dissimilatory nitrate reduction" as some of the microorganisms assimilate ammonium for synthesis of proteins and amino acid.

5. Denitrification: This is the reverse process of nitrification. During denitrification nitrates are reduced to nitrites and then to nitrogen gas and ammonia. Thus, reduction of nitrates to gaseous nitrogen by microorganisms in a series of biochemical reactions is called "denitrification". The process is wasteful as available nitrogen in soil is lost to atmosphere. The overall process of denitrification is as follows:

Nitrate -----> Nitrite ----> Nitric Oxide ----> Nitrous Oxide ------> Nitrogen gas

This process also called dissimilatory nitrate reduction as nitrate nitrogen is completely lost into atmospheric air. In the soils with high organic matter and anaerobic soil conditions (waterlogged or ill-drained) rate of denitrification is more. Thus, rice / paddy fields are more prone to denitrification.

The most important denitrifying bacteria are *Thiobacillus denitrificans, Micrococcus denitrificans,* and species of *Pseudomonas, Bacillus, Achromobacter, Serrtatia paracoccus* etc.

Denitrification leads to the loss of nitrogen (nitrate nitrogen) from the soil which results into the depletion of an essential nutrient for plant growth and therefore, it is an undesirable process / reaction from the soil fertility and agricultural productivity. Although, denitrification is an undesirable reaction from agricultural productivity, but it is of major ecological importance since, without denitrification the supply of nitrogen including N_2 of the atmosphere, would have not got depleted and NO_3 (which are toxic) would have accumulated in the soil and water.

Mechanism of Biological Nitrogen Fixation

A relatively small amount of ammonia is produced by lightning. Some ammonia also is produced industrially by the Haber-Bosch process, using an iron-based catalyst, very high pressures and fairly high temperature. But the major conversion of N_2 into ammonia, and thence into proteins, is achieved by microorganisms in the process called nitrogen fixation (or dinitrogen fixation). The total biological nitrogen fixation is estimated to be twice as much as the total nitrogen fixation by non-biological processes.

Biological nitrogen fixation can be represented by the following equation, in which two moles of ammonia are produced from one mole of nitrogen gas, at the expense of 16 moles of ATP and a supply of electrons and protons (hydrogen ions):

$$N_2 + 8H^+ + 8e^- + 16\ ATP = 2NH_3 + H_2 + 16ADP + 16\ Pi$$

This reaction is performed exclusively by prokaryotes (the bacteria and related organisms), using an enzyme complex termed **nitrogenase.**

The nitrogenase system is present only in prokaryotes. It consists of two dissociating protein components. Component I (also called enzyme I, molybdoferredoxin, azofermo and Mo-Fe protein), contains Mo, Fe and S--. Component II, (also called enzyme II, azoferrodoxin, azofer and Fe protein), contains Fe and S. Each component by itself has no activity of its own. Combination of the two inactive components reconstitutes nitrogenase activity.

The two components interact as a complex in vivo, and one component is required for the stability of the other. The two inactive components have similar physical and chemical properties. Nitrogenase is the diagnostic enzyme for nitrogen-fixing organisms; it consists of two proteins:

(i) **Dinitrogenase:** This component reduces nitrogen (N_2) to ammonia (NH_3). It contains an iron-molybdenum cofactor (abbreviated FeMoCo) that accepts electrons from dinitrogenase reductase.

(ii) **Dinitrogenase reductase**: This component transfers electrons to dinitrogenase. It contains an iron atom that is involved in the redox chemistry. A noteworthy limitation: dinitrogenase reductase is irreversibly inactivated by oxygen. As a result, nitrogen fixation cannot be done in the presence of oxygen.

Symbiotic Nitrogen Fixation Microorganisms		**Host plants**		**Location**	**Isolated**
Large group	Genera	Plant group	Tissue	Inside or outside plant cell	
Bacteria (-Proteobacteria)	Rhizobium, Bradyrhizobium, Azorhizobium	Legumes and Parasponia	Nodule (induced)	Inside	Yes
Actinomycetes	Frankia	Betulaceae and 8 family(trees)	Nodule (induced)	Inside	Yes
Cyanobacteria	Nostoc	Bryophytes (Antheros *etc.*)	Leaf cavity	Outside	Yes
	Nostoc (Anabaena?)	Pteridophyte (Azolla)	Leaf cavity	Outside	No
	Nostoc	Cycadopha (Cycas,Macrozamia *etc.*)	Collaroid root	Outside	Yes
	Nostoc	Angiosperm (Gunnera)	Gland tissue	Inside	Yes

Alternate Nirogenase

Type	**Metals of dinitrogen reductase**	**Activity (NH3 formed nmol/min/ mg enzyme)**	**Organisms**	**Gene code**
1(Mo-nitrogenase)	Fe,Mo	1000	All	nif
2(V-nitrogenase)	Fe,V	1000~300	Clostridium,Azotobacter, Anabaena	vnf
3(nitrogenase-3)	Fe	<50	Azotobacter, Rhodobacter, Rhodopseudomonas (photosynthetic)	anf

Remark: Acitivities of pure enzyme. Alternate nitrogenases can reduce acetylene up to ethane.

Mechanism escaping O_2 damage

Avoid (Escape)	Move to lower O_2 pressure	Azospirillum
	High Oxygen uptake by respiration	Azotobacter, Derxia
O_2 scavenging	Hemoglobin combines efficiently with O_2	In nodules of rhizobia-legume symbiosis, Casuarina-Frankia symbiosis
Protect	Some proteins protect nitrogenase under O_2 exposure	Azotobacter
	Nitrogenase in heterocysts (Heterocysts do not have photosynthetic ability, do not evolve O_2)	Cyanobacteria
Separation in space	in Vesicles in Frankia-nodules symbiosis	Frankia
	In colony, differentiate into photosynthetic filamants and nitrogen fixing filaments	Marine Scenedesmium
Separation in time	Photosynthesis daytime, and nitrogen fixation at night.	Uni-cellular Cyanobacteria

Generally, Nitrogen Fixation is active under low (<0.1% v/v) O_2 pressure

Mechanism of Nitrogenase Action

In most organisms, the physiologically functional reductant of nitrogenase is ferredoxin. Other natural reductants include flavodoxin and NADPH. Artificial reductants include $Na_2S_2O_4$ and reduced methyl viologen. Reduced ferredoxin, the electron donor, reduces the Fe protein of nitrogenase.

Fe Protein (Oxidized) + e^- --> Fe Protein (reduced)

The reduction of N_2 to NH_4 is exothermic. Yet nitrogen fixation requires energy in the form of adenosine 5′-triphosphate (ATP), because of a high activation energy.

The Fe protein of nitrogenase specifically binds to Mg ATP and lowers its redox potential. A complex containing both Fe and Mo-Fe proteins and Mg ATP is assembled.

The Fe protein transfers an electron to the Mo-Fe protein. This results in the oxidation of the Fe protein and the reduction of the Mo-Fe protein. This reaction is coupled to ATP ---> ADP hydrolysis. ATP is not hydrolyzed to A TP until the Fe protein transfer an electron to the Mo-Fe protein. There are 12 or more ATP hydrolyzed for each N_2 reduced, or 4 ATP per pair of electrons transferred to the Mo- Fe protein or to the substrate. Thus, there appear to be 2 ATPs hydrolyzed for each electron transferred. In vivo and in vitro ATP requirements are not necessarily the same. The reduced Mo-Fe protein can in turn reduce the substrate. A number of substrates other than N_2 can be reduced by nitrogenase. Both Fe and Mo-Fe proteins are required for all these reductions, which are coupled to Mg ATP hydrolysis.

Hydrogen Evolution

Nitrogenase forms molecular hydrogen, if no suitable electron acceptor is available. Hydrogen is evolved by many nitrogen fixing bacteria during nitrogen fixation. No hydrogen is, however, evolved when they are grown on NH^{+}_{4}, because nitrogenase synthesis is repressed. Nitrogenase itself can be responsible for H_2, evolution.

Nitrogenase receives electrons during nitrogen fixation, and transfers them to the dinitrogen molecule. three electrons (e^-) associate with each atom of nitrogen. Three protons (H^+) are therefore withdrawn from the intracellular medium to neutralize, the negative charge of the electrons. Thus, each dinitrogen molecule yields two molecules of ammonia.

Electron transfer from nitrogenase to molecular nitrogen has a side reaction which is energetically wasteful. In the absence of N_2 and in the presence of ATP and electrons, the electrons combine with protons to yield H_2, ATP is hydrolyzed during hydrogen evolution. Even when gaseous nitrogen is available, some electrons are evolved as H_2. The evolved hydrogen is of no apparent use to the plant or the bacteria. Energy derived from metabolism is used in separating free protons (H^+) and electrons (e^-) of $H_2$for use in driving the nitrogen fixation reaction. When the protons and electrons recombine to form H^+, the energy is wasted.

H_2 is a competitive inhibiter of N_2 fixation. Some strains of Rhizobium carry the hup gene in Fe Protein in the plasmid. Such strains are able to synthesize hydrogen oxidizing uptake hydrogenase. This enzyme converts molecular hydrogen back to electrons and protons, which arc then reused by nitrogenase. The increased bacterial efficiency would enable the host plant to direct less energy for supporting the bacteria, and more for increasing seed yield. Wild type Rhizobium strains capable of oxidizing H_2 through hydrogenase appear to be more efficient in nitrogen fixation than strains without hydrogenase.

Factors limiting biological nitrogen fixation

1. Edaphic Factors
2. Climatic Factors
3. Biotic Factors

Interactions between the microsymbiont and the plant are complicated by edaphic, climatic, and management factors. A legume-Rhizobium symbiosis might perform well in a loamy soil but not in a sandy soil, in the subhumid region but not in the Sahel, or under tillage but not in no-till plots. These factors affect either the microsymbiont, the host-plant, or both.

1. Edaphic Factors

Edaphic factors relate to the soil. The six main edaphic factors limiting biological nitrogen fixation are: excessive soil moisture, drought, soil acidity, P deficiency,excess mineral N, and deficiency of Ca, MO, CO and B.

Excessive moisture and waterlogging prevent the development of root hair and sites of nodulation, and interfere with a normal diffusion of O_2 in the root system of plants. Sesbania rostrata and Aeschynomene sp. can actively fix N_2 under these conditions because they are located on the plant stems, rather than on the roots.

Drought reduces the number of rhizobia in soils, and inhibits nodulation and N_2 fixation. Prolonged drought will promote nodule decay. Deep-rooted legumes exploiting moisture in lower soil layers can continue fixing N_2 when the soil is drying. Mycorrhizal infection has also been found to improve tolerance of plants to drought (*e.g.*, Acacia auriculiformis inoculated with the ectomycorrhizal Baletus suillus).Mycorrhiza are symbiotic associations between fungi and plant roots. Some mycorrhizal fungi develop exclusively outside the roots; these are called ectomycorrhiza (*e.g.*, Baletus suillus).Others, called endomycorrhiza, grow inside the roots with their vesicles and arbuscules inside the roots and with their fungal filaments extended outside (*e.g.*, Glamus sp.).These are the vesicular-arbuscular mycorrhiza, usually referred to as VAM.

Soil acidity and related problems of Ca deficiency and aluminum and manganese toxicity adversely affect nodulation, N_2 fixation and plant growth . Research work on the identification of symbioses adapted to acid soil should focus on the host plant, because effective rhizobia adapted to- soil acidity can be found naturally and can be produced through genetic manipulations.

Phosphorus deficiency is commonplace in tropical Africa and reduces nodulation, N_2 fixation and plant growth. Identification of plant species adapted to low-P soils is a good strategy to overcome this soil constraint. The role of mycorrhizal fungi in increasing plant P uptake with beneficial effects on N_2 fixation has been reported. Dual inoculation with effective rhizobia and mycorrhizal fungi shows synergistic effects on nodulation and N_2 fixation in low P soils*. The use of local rock phosphate has been recommended, particularly in acid soils, as an inexpensive source of P. The addition of P-solubilizing microorganisms, particularly of the general Psemdamaias, Bacillus, Penicillium, and Aspergillus can solubilize rock phosphate and organically bound soil P (which constitutes 95 - 99% of the total phosphate in soils). However, the use of these microorganisms is not widespread. Some reports show nodulation response to K under field conditions. However, other investigators consider the K effect to be indirect, acting through the physiology of the plant.

* Trees are usually infected by mycorrhizal fungi in natural ecosystems in the tropics. The significance of this symbiosis in nature should be better recognised.

Mineral N inhibits the Rhizobium infection process and also inhibits N_2 fixation. The former problem probably results from impairment of the recognition mechanisms by nitrates, while the latter is probably due to diversion of photosynthates toward assimilation of nitrates. Some strains of Rhizobium, and particularly stem-nodulating Azarhizobium caulinodans, fix N_2 actively even when plants are growing in high-N soils (*e.g.*, in the presence of 200 kg fertilizer N ha^{-1}) . Application of large quantities of fertilizer N inhibits N_2 fixation, but low doses (<30 kg N ha^{-1}) of fertilizer N can stimulate early growth of legumes and increase their overall N_2 fixation. The amount of this starter N must be defined in relation to available soil N.

Various microelements (Cu, Mo, Co, B) are necessary for N_2 fixation. Some of these are components of nitrogenase for example Mo.

2. Climatic Factors

The two important climatic determinants affecting BNF are temperature and light.

Extreme temperatures affect N_2 fixation adversely. This is easy to understand because N_2 fixation is an enzymatic process. However, there are differences between symbiotic systems in their ability to tolerate high (>35°C) and low (<25°C) temperatures.

The **availability of light** regulates photosynthesis, upon which biological nitrogen fixation depends. This is demonstrated by diurnal variations in nitrogenase activity. A very few plants can grow and fix N_2 under shade (*e.g.*, Flemingia congesta under plantain canopy). In alley farming if hedgerows are not weeded, or if trees are planted with food crops like cassava, their nitrogen fixation and growth will be reduced due to shading. Early growth of legume trees is slow and they cannot compete successfully for light.

3. Biotic Factors

Among biotic factors, the absence of the required rhizobia species constitute the major constraint in the nitrogen fixation process. The other limiting biotic factors could be:

- Excessive defoliation of host plant,
- Crop competition,
- Insects and nematodes

Inoculation of Legumes

If specific and effective rhizobia are absent in a soil, or if they are present in low numbers, it is necessary to introduce the rhizobia in that soil to ensure proper nodulation and nitrogen fixation. This is called inoculation. If specific and effective rhizobia are present in a sufficient number, there will be no need to inoculate the legume. In agrisystems, whenever one is not sure of the presence and effectiveness of the native rhizobia, it could be necessary to inoculate the legume with an adequate strain of rhizobia.

A more accurate relative effectiveness trial will provide more precise information. The trial, in a simple term, consists of growing the legume with and without fertilizer N while controlling all other limiting factors. The relative effectiveness ratio (RE) is then calculated. RE is defined as: dry weight of unfertilized plants × 100/dry weight of fertilized plants. If the value of RE is more than 5, the inoculation is not required.

When the rhizobia in a soil are infective (*i.e.*, capable of colonizing and nodulating a legume) but poorly effective, they constitute a barrier to the successful exploitation of Rhizobium inoculants. Introduced rhizobia must therefore be more aggressive and competitive as nodulators than the native strains. Inoculant rhizobia usually persist in the soil for long periods, particularly when the host is cultivated frequently or is permanent. Persistence of a strain is desirable because it obviates the need for inoculation in subsequent years, assuming inoculant strains maintain their original effectiveness.

Inoculation with rhizobia is usually recommended for newly introduced legumes. Most positive responses to inoculation are confined to crops which have specific requirements for Rhizobium, (*e.g.*, Leucaena leucocephala, American varieties of soybean). Indigenous legumes seldom respond to inoculation with introduced rhizobia because they nodulate with resident strains, even if these native rhizobia are not the most effective ones.

Inoculation with rhizobia should be considered as an exceptional farming practice rather than the rule. In Australia and the USA, inoculation has played a vital role in legume production. But in developing countries, the practice is not widespread. The major drawback to inoculation technology is the wide variability in yield responses in time and space for a given Rhizobium-legume symbiosis. Responses can vary from no response, and sometimes negative responses, to positive yield increases. Response to inoculation with a strain of Rhizobium vary with sites, legume cultivars, and the form of inoculant. Changes in climate, such as Africa's long droughts in recent years, and management factors including cropping systems and inoculant handling will also introduce variability in response to inoculation. Local rhizobia are not necessarily better inoculants than exotic strains.

All these considerations call for a substantial research support system capable of defining the most appropriate inoculants and procedures for each site and probably for each cropping season as well. The use of freely nodulating legumes will be much easier in this respect.

- Defoliation, Crop Competition, and Pests
- Defoliation (*e.g.*, pruning and lopping) decreases the photosynthetic ability of legumes. It impairs N_2 fixation and can lead to nodule decay. For perennial legumes, nodule decay sheds a high number of rhizobia in the root zone. When new roots develop in subsequent vegetative cycles, nodulation of the legume is expected to improve. Scientists at IITA have observed that uninoculated Leucaena leucocephala nodulated very sparsely the first year and showed nitrogen deficiency symptoms. After a number of years nodulation improved and N deficiency symptoms disappeared.
- Intercropping legumes with non-leguminous crops can result in competition for water and nutrients. This competition can affect N_2 fixation negatively. However, it has been shown that when mineral N is depleted in the root zone of the legume component by the non-leguminous intercrops, N_2 fixation of legumes may be promoted.
- Insects and nematodes have also been reported to interfere with nodule formation, development, and functions.

Nitrogen Fixation

Fixation of elemental nitrogen in the atmosphere by the microorganisms through a reductive process into ammonia is called as biological nitrogen fixation

(BNF). A variety of prokaryotic organisms have the ability to reduce the atmospheric nitrogen. BNF accounts for about 70% of the total N fixed in the biosphere. The ability to reduce atmosphere N is restricted only to bacteria, which are belonging to the diverse groups. Rhizobia are the first group of organisms realized for its potential of nitrogen fixation.

The process of N_2 fixation is mediated by the enzyme, called nitrogenase (which mediates the reduction of N_2 to ammonia). The overall reaction is as follows.

$$N_2 + 3H_2 \xrightarrow{ATP} 2\,NH_3$$

Some of the nitrogen fixing organisms live independently of other organisms - the so-called free-living nitrogen-fixing bacteria and others live in intimate symbiotic associations with plants or with other organisms. The examples for free living or non-symbiotic N_2 fixers are *Azotobacter, Beijerinckia, Klebsiella,* Cyanobacteria, *Clostridium* and *Desulfovibrio*. The well known examples for symbioticN_2 fixers are Rhizobium, Bradyrhizobium and Azorhizobium with Legume plants; Frankia with Casuarina; Anabaena (Cyanobacteria) with Azolla (water fern).

Nitrogen fixing bacteria are classified according to their mode of fixation.

3. Free living N fixers – capable of fixing mol. N_2 to cellular N independently of other living organism.
4. Associative N fixers
5. Endophytic N fixers
6. Symbiotic N fixers

Rhizobium is predominant symbiotic N_2 fixing bacterium. Boussingault showed that leguminous plant can fix atmosphere N_2. Then Hellriegel and Wilfarth – proved that N_2 is fixed by certain bacteria living in root nodules of leguminous plants. Latter isolated in pure culture by Beijerinck. Winogradsky isolated Clostridium pasteurianum, which is an anaerobic N_2 fixer. Beijerinck isolated Azotobacter as a free living aerobic N_2 fixing organisms.

Cross inoculation groups of rhizobium (CIG)

It (CIG) refers the groups of leguminous plants that will develop effective nodules when inoculated with the rhizobia obtained from the nodules from any member of that legume group.

I. Rhizobium

1. Rhizobium leguminosarum

   CIG Host it can nodulate

   bv. *viceae* Peas, lenfils, vicia bv. phaseoli Bean *Phaseolus* spp bv. *trifoli* Clover *Trifolium* spp
2. R. *meliloti* Alfalfa Alfalfa, clover, fenugreek
3. R. *loti* Lotus Trifoli, lupine,

4. R. *fredii* Soybean
5. R. spp Cowpea group *Vigna*, Arachis, Cajanus, Dolichus, Sesbania, Acacia, Prosopis, green gram and blackgram
6. R. sp Chickpea group Chickpea

II. *Bradyrhizobium*

B. japonicum Soybean

B. spp Cowpea group plants

III. *Azorhizobium* - Stem nodulating – one. Nodulates Sesbania rostrata.

IV. *Photorhizobia* - Nodulants *Aeschynomene* sp.

V. *Sinorhizobium* - fast growing soybean nodulator

I. Biological nitrogen fixation

Free living nitrogen fixers

Azotobacter – Aerobic *Beijerinckia*

Clostridium – Anaerobic

Cyanobacteria (Blue green algae) *etc.,*

II. Associative symbiotic nitrogen fixer

Azospirillum

Herbaspirillum

III. Endophytic nitrogen fixer

Gluconacetobacter diazotrophicus

IV. Symbiotic nitrogen fixers

Rhizobium (*Rhizobium* – legume association)

Bradyrhizobium (*Bradyrhizobium* – soybean association)

Azorhizobium (*Azorhizobium- Sesbania rostrata* association)

Anabaena azollae (Azolla – *Anabaena association*)

Frankia (*Frankia* – Casuarina association)

Species of Azospirillum

A. lipoferum

A. brasilense

A. amazonense

A. halopareferans

A. irkense

Species of *Azotobacter*

A. chroococcum

A. vinelandii

A. beijerinkii

A. paspali

A. agilis

A. insignis

A. macrocytogens

Important genera of blue green algae

Anabaena, Nostoc, Cylindrospermum, Rivularia, Oscillatoria, Plectonema, Aphanothece, Lyngbya, Scytonema, Calotrhix etc.,

Species of Azolla

A. pinnata

A. filiculoides

A. microphylla

A. caroliniana

A. mexicana

A. nilotica

Nitrogen fixation

Rhizobium is the most well known species of a group of bacteria that acts as the primary symbiotic fixer of nitrogen. These bacteria can infect the roots of leguminous plants, leading to the formation of lumps or nodules where the nitrogen fixation takes place. The bacterium's enzyme system supplies a constant source of reduced nitrogen to the host plant and the plant furnishes nutrients and energy for the activities of the bacterium. About 90% of legumes can become nodulated.

In the soil the bacteria are free living and motile, feeding on the remains of dead organisms. Free living rhizobia cannot fix nitrogen and they have a different shape from the bacteria found in root nodules. They are regular in structure, appearing as straight rods; in root nodules the nitrogen-fixing form exists as irregular cells called bacteroids which are often club and Y-shaped.

Root nodule formation

Sets of genes in the bacteria control different aspects of the nodulation process. One Rhizobium strain can infect certain species of legumes but not others *e.g.* the pea is the host plant to Rhizobium leguminosarum biovar viciae, whereas clover acts as host to R. leguminosarum biovar trifolii. Specificity genes determine which Rhizobium strain infects which legume. Even if a strain isable to infect a legume, the nodules formed may not be able to fix nitrogen. Such rhizobia aretermed ineffective.

Effective strains induce nitrogen-fixing nodules. Effectiveness is governed by adifferent set of genes in the bacteria from the specificity genes. Nod genes direct the variousstages of nodulation. The initial interaction between the host plant and free-living rhizobia is therelease of a variety of chemicals by the root cells into the soil. Some of these encourage the growthof the bacterial population in the area around the roots (the rhizosphere).

Legume effect: The moment when the plant is raised, the rhizobia population build up is very high and this effect is not noticed when non legume is raised. Previously it was thought that homoserine copounds were the triggering this effect. Later it was proved that the signal compounds in the root exudates containing are responsible. Signal compounds are not secreted by all plants, only legume alone are producing and they fall into four categories

- Flavones
- Flavanoids
- Isoflavones
- Chalcones

The Signal Compounds are

- Low molecular weight compounds
- Secondary metabolites
- Produced in micromolar quantities
- General structure has benzene ring
- Recognizes Rhizobia
- *e.g.* Apigenin, Quercetin, Leutinol, Niranginin, Chrysin, *etc.*

Flavonoids secreted by the root cells activate the nod genes in the bacteria, which then induce nodule formation. The whole nodulation process is regulated by highly complex chemical communications between the plant and the bacteria. Lectins are low molecular carbohydrate binding protein present in almost in all plants and microbes donot produce lectin. Cell wall of root hair contain lectin which helps to attach with exoployscharide of rhizobia and there by cause lectin binding reaction. Without lectin cementing could not be possible and it also helps in orientation of rhizobia on root hair. Once binding occurs to the root hair, the bacteria excrete nod factors. These stimulate the hair to curl. Rhizobia then invade the root through the hair tip where they induce the formation of an infection thread. This thread is constructed by the root cells and not the bacteria and is formed only in response to infection. The infection thread grows through the root hair cells and penetrates other root cells nearby often with branching of the thread. The bacteria multiply within the expanding network of tubes, continuing to produce nod factors, which stimulate the root cells to proliferate, eventually forming a root nodule. Within a week of infection small nodules are visible to the naked eye. Each root nodule is packed with thousands of living Rhizobium bacteria, most of which are in the misshapen form known as bacteroids. Portions of plant cell membrane surround the bacteroids. These structures, known as symbiosomes, which may contain several bacteroids or just one, are where the nitrogen fixation takes place.

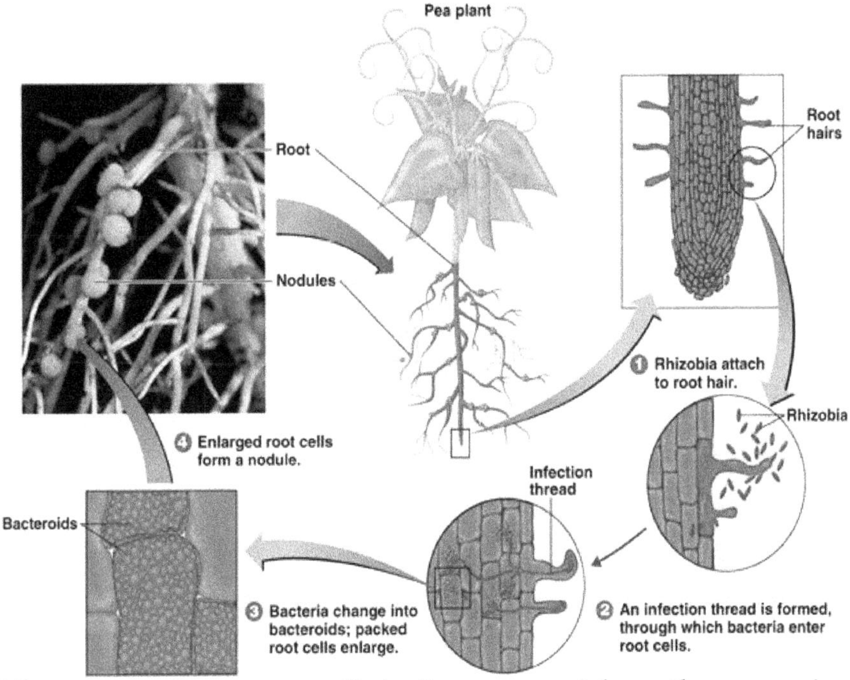

Nitrogenase: An enzyme called nitrogenase catalyses the conversion of nitrogen gas to ammonia in nitrogen-fixing organisms. In legumes it only occurs within the bacteroids. The reaction requires hydrogen as well as energy from ATP. The nitrogenase complex is sensitive to oxygen, becoming inactivated when exposed to it. This is not a problem with free living, anaerobic nitrogen-bacteria such as Clostridium. Free-living aerobic bacteria have a variety of different mechanisms for protecting the nitrogenase complex, including high rates of metabolism and physical barriers.

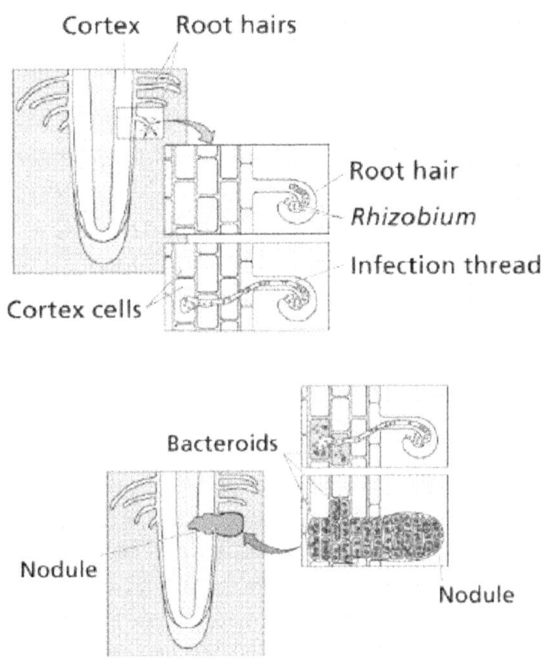

Rhizobium controls oxygen levels in the nodule with leghaemoglobin. This red, iron-containing protein has a similar function to that of haemoglobin; binding to oxygen. This provides sufficient oxygen for the metabolic functions of the bacteroids but prevents the accumulation of free oxygen that would destroy the activity of nitrogenase. It is believed that leghaemoglobin is formed through the interaction of the plant and the rhizobia as neither can produce it alone.

Leghaemoglobin: Synthesized by plant in the nodule tissue when infection thread transform to bacteroids Leghemoglobin synthesis switches on when bacteroid formation initiates. It helps to mainatain redox potential of nodule, helps in absorbtion of oxygen and not permitting to destroy nitrogenase, helps to supply nutrients to bacteroids.

Important genera

Rhizobium	Bradyrhizobium
Allorhizobium	Mesorhizobium
Sinorhizobium	Azorhizobium

There two types of nodules

Determinate - Definite shape different size & shape

Indeterminate - Nodule dev. Sieze at one point continuous process

e.g. Goundnut Bengal gram

FRANKIA

Productivity in forest ecosystems is often limited by low levels of available nitrogen, which results from the relatively slow rates of mineralization of organic N in the soil. Nitrogen deficiency is always pronounced in nursery seedling stage. Even though application of mineral nitrogen can compensate and provide the deficiency, besides its cost factor, the undesirable environmental reasons are important factors for consideration. Such factors, coupled with projection of increased costs of fertilizer production, have been largely responsible for stimulating interest in the development of forest management techniques in which biological N-fixation provides a more gradual input of N to meet the N-requirement of the crop.

The term "actinorhiza" refers both to the filamentous bacteria Frankia, an actinomycete, and to the root location of the nitrogen fixing nodules. Actinorhizal plants are classified in four subclasses, eight Families and 25 genera comprising more than 220 species. Basic studies in complex interactions between microorganisms and actinorhizal plants will help us also to increase our knowledge about plant growth regulation and their interactions with microbes in nature. Frankia is a nitrogen fixing actinomycete, which can fix atmospheric nitrogen in symbiotic association. Frankia infects its host plant forming a nodule on the root system. In the symbiosis, Frankia provides the plant with a source of nitrogen, and in return the plant provides the Frankia with a carbon source. Two pathways for root infection have been described for compatible Frankia interactions: Root hair infection, as in Alnus or intercellular penetration, as in Discaria.

Losses of N by non biological ways

Leaching

20 to 50% of fertilizer N. The most striking loss of N in rice soils where more than half of the fertilizer N applied get lost through leaching.

Volatalization

Another factor is the volatalizaiotn of ammonia in soil 5-20%. Fixation of ammonium in soils is the minor contributory factor to overall loss of N_2 available for plant growth. Such losses of N by physical causes and by nitrification and denitirfication process can be controlled by the application of certain chemicals. Some chemicals have been designed to control the rate of release of nutrient from nitrogenous fertilizers, while others retard nitrification in soil by controlling the activity of nitrifying bacteria.

a. Controlled release fertilizers

Urea from isobutyeldene diurea

Crotonilidene diurea

S coated urea

b. Nitrification inhibitors

These are substituted with pyridines, pyrimidines, anilines and isothiocyanates,

Examples

1. 2 chloro 6 (tricholormethyl) – pyridine – (N serve)

2. 2 amino 4 chloro 6 methyl pyridine –(AM.)

N serve inhibits the growth of Nitrosomonas europea and N. agilis.

The seeds of neem conain lipid associates act as nitrification inhibitors and there by increases the efficiency of urea fertilizers.

Fertilizers, sparingly soluble in water can regulate the release of N from fertilizers

Ammonia assimilation

N_2 fixation results in NH_4 formation which reacts with organic acids and form amino acids which is mediated by ammonia assimilating enzyme.

GS – Glutamine synthetase

GOGAT – Glutamate synthese

GDH – Glutamate dehydrogenase

Genetics

Nif genes are responsible for N_2 fixation.

Nif genes are 22, which are located in 7 or 8 clusters.

Carbon Cycle The most important element in the biological realm and substance that serve as the cornerstone of the cell structure is carbon. It constituents about 40-50% of all living organisms, yet the ultimate source is the CO_2 that exist in a perennially short supply, only 0.03% of the earth's atmosphere, which undergo a cyclic change from an oxidized to reduced state. The concentration of carbon in living matter (18%) is almost 100 times greater than its concentration in the earth (0.19%). Hence, living things extract carbon from their non-living environment. For life to continue, this carbon must be recycled. Carbon exists in the nonliving environment as: carbon dioxide (CO_2) in the atmosphere and dissolved in water (forming HCO_3–) carbonate rocks (limestone and coral = $CaCO_3$) deposits of coal, petroleum, and natural gas derived from once-living things dead organic matter, *e.g.*, humus in the soil. Carbon enters the biotic world through the action of autotrophs: primarily photoautotrophs, like plants and algae, that use the energy of light to convert carbon dioxide to organic matter. Links to photosynthesis and to a small extent, chemoautotrophs - bacteria and Archaeans that do the same but use the energy derived from an oxidation of molecules in their substrate.

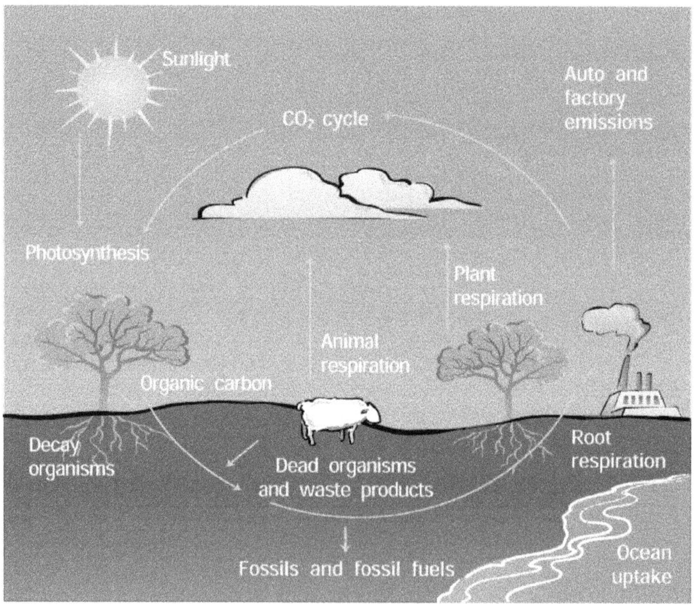

The carbon cycle revolves about CO_2 and its fixation and regeneration. The green plants utilize CO_2 as their sole carbon source, and the carbonaceous matter synthesized serves to supply carbon to other heterotrophic organisms and animals. Upon the death of plants and animals, microbes assume a dominant role in carbon cycle. The dead tissues are degraded and transformed into microbial cells and humus or soil organic fraction. Further decomposition of these materials leads to the production of CO_2 (producing CO_2 if oxygen is present, methane - CH_4 if oxygen is not present) and once again it is recycled.

The uptake and return of CO_2 are not in balance. The carbon dioxide content of the atmosphere is gradually and steadily increasing. This increase is surely "anthropogenic" that is, caused by human activities: burning fossil fuels (coal, oil, natural gas), clearing and burning of forests, especially in the tropics. In recent decades, large areas of the Amazon rain forest have been cleared for agriculture and cattle grazing. Curiously, the increase in atmospheric CO_2 is only about one-half of what would have been expected from the amount of fossil fuel consumption and forest burning.

Carbon assimilation

The process of converting substrate to protoplasmic carbon is known as assimilation. Under aerobic conditions 20-40% of the substrate carbon is assimilated, the remainder is released as CO_2. Fungi are more efficient, in their metabolism, since they convert carbon into cell carbon as filaments and releases less of CO_2. Around 30 - 40 % is used to form new mycelium during the decomposition. Compared to fungi, bacteria are less efficient.

Mineralization

Conversion of organic carbon substance to inorganic form of carbon is known as carbon mineralization.

Immobilization

Assimilation of nutrients by the microorganisms called immobilization, which reduces the quantity of available nutrients to the plants.

Composition of Organic Matter

Soil organic matter plays important role in the maintenance and improvement of soil properties. It is a dynamic material and is one of the major sources of nutrient elements for plants. Soil organic matter is derived to a large extent from residues and remains of the plants together with the small quantities of animal remains, excreta, and microbial tissues. Soil organic matter is composed of three major components *i.e.* plants residues, animal remain and dead remains of microorganisms. Various organic compounds are made up of complex carbohydrates, (Cellulose, hemicellulose, starch) simple sugars, lignins, pectins, gums, mucilages, proteins, fats, oils, waxes, resins, alcohols, organic acids, phenols *etc.* and other products. All these compounds constituting the soil organic matter can be categorized in the following way.

Soil Organic matter

Soil organic matter comprises residues of plant and animals and these compounds occur in soil in close combination with inorganic substances. Animals and plant residues are made up of complex carbohydrates, simple sugars, starch, cellulose, hemicellulose, pectins, gums, mucilage, proteins fats, oils, waxes, resins, alcohols, aldehydes, ketones, organic acids, lignin, phenols, tannins, hydrocarbons, alkaloids, pigments *etc.*

The organic matter subjected to microbial decay in the soil comes from several sources like plant remains, forest litter, incorporation of plant tissues, animal tissues, and excretory products. The chemistry of organic matter is clearly very complex, and investigations of the transformations and the responsible organisms have therefore been extremely interesting. The organic constituents of the plants are commonly divided into six categories.

a. **Cellulose** - Most abundant 15-60% of the dry weight

b. **Hemicellulose** - 10-30% of the plant dry weight

c. **Lignin** - 5 - 30 % of the plant dry weight

d. **Water soluble fraction** - 5 - 30 %, includes simple sugars and amino acids

e. **Ether and alcohol soluble constituents**, a fraction containing fat, oils, waxes, resins and pigments

f. **Proteins** - As the plant ages, the content of water soluble constituents, proteins and minerals decreases. Hence % of abundance of cellulose, hemicellulose and lignin rises. The relationship between organic matter and plant growth may be direct or indirect.

- Organic matter is a natural substrate for saprophytic microorganisms and provides nutrition to plants indirectly through the activity of soil microorganisms
- It is essential for the formation of soil aggregates and hence soil structure which ultimately determines the soil aeration and rooting habit of plants
- Organic matter helps in the conservation of soil nutrients by preventing erosion and surface run off of nutrients.

Organic Matter (Undecomposed)

A. Organic

Nitrogenous:

1. Water Soluble *e.g.* Nitrates, ammonical compounds, amides, amino acids *etc.*
2. Insoluble *e.g.* Proteins nucleoproteins, peptides, alkaloids purines, pyridines chitin *etc.*

Non Nitrogenous:

- Carbohydrates *e.g.* Sugars, starch, hemicellulose, gums, mucilage, pectins, *etc.*
- Micellaneous: *e.g.* Lignin, tannins, organic acid, *etc.*
- Ether Solube: *e.g.* Fats, oils, wax *etc.*

B. Inorganic

The organic complex / matter in the soil is, therefore made up of a large number of substances of widely different chemical composition and the amount of each substance varies with the type, nature and age of plants. For example cellulose in a young plant is only half of the mature plants; water-soluble organic substances in young plants are nearly double to that of older plants. Among the plant residues, leguminous plants are rich in proteins than the non-leguminous plants. Grasses and cereal straws contain greater amount of cellulose, lignin, hemicelluloses than the legumes and as the plant gets older the proportion of cellulose, hemicelluloses and lignin gets increased. Plant residues contain 15-60% cellulose, 10-30 % hemicelluslose, 5-30% lignin, 2-15 % protein and 10% sugars, amino acids and organic acids. These differences in composition of various plant and animal residues have great significance on the rate of organic matter decomposition in general and of nitrification and humification (humus formation) in particular. The end products of decomposition are CO_2, H_2O, NO_3, SO_4, CH_4, NH_4, and H_2S depending on the availability of air.

Organic matter decomposition

The soil microorganisms play an important role in the decomposition of organic matter. Bacteria are the dominant group in the decomposition; mostly heterotrophic organisms (use energy from organic sources such as sugars, starch, cellulose and protein) are actively involved. Autotrophic organisms occupy a small portion of the biomass in soil are not directly involved in organic matter decomposition. Soil fungi are mostly heterotrophs and use organic residues easily.

Soil algae contribute a small amount of organic matter through their biomass, but they do not have any active role in organic matter decomposition.

During the decomposition of organic matter three separate simultaneous processes can be distinguished. The important changes during decomposition are:

- Plant and animal tissues constituents disappear under the influence of enzymes.
- Synthesis of new microbial cells so that proteins, polysaccharides and nucleic acids of bacteria and fungi appear.
- Third, certain end products of the breakdown are excreted into surroundings to accumulate or to be further metabolized.

Genera of microorganisms capable of utilizing different components of organic matter as reported by several workers: F-Fungi; B-Bacteria; A-Actinomycetes

Nature of substrate in organic matter		Genera of microorganisms
Cellulose	F	*Alternaria, Aspergillus, Chaetomium, Coprinus, Fomes, Fusarium, Myrothecium, Penicillum, Polyporus, Rhizoctonia, Rhizopus, Trametes, Trichoderma, Trichothecium, Verticillium, Zygorynchus*
	B	*Achrombacter, Angiococcus, Bacillus, Cellfalcicula, Cellulomonas, Cellvibrio, Clostrium, Cytophaga, Polyangium, Pseudomonas, Sorangium, Sporocytophaga, Vibrio*
	A	*Micromonospora, Nocardia, Streptomyces, Streptosporangium*

Nature of substrate in organic matter		Genera of microorganisms
Hemicellulose	F	*Alternaria, Fusarium, Trichothecium, Aspergillus, Rhizopous, Zygorynchus, Chateomium, Helminthosporium, Penicillium, Coriolus, Fomes, Polyporus*
	B	*Bacillus, Achromobacter, Pseudomonas, Cytophaga, Sporocytophaga, Lactobacillus, Vibrio*
	A	*Streptomyces*
Lignin	F	*Clavaria, Clitocybe, Collybia, Flammula Hypholoma, Lepiota, Mycena, Pholiota, Arthrobotrys, Cephalosporium, Humicola*
	B	*Pseudomonas, Flavobacterium*

Anaerobic decay / decomposition

The main products of aerobic carbon mineralization are CO_2, water, cells and humus components. In the absence of O_2 organic carbon is incompletely metabolized, intermediary substances accumulate, and abundant quantities of CH_4 and smaller amounts of H_2 are evolved. Energy yield during anaerobic fermentation is low, resulting in fewer microbial cells per unit of organic carbon degraded. Consequently, organic matter breakdown is consistently slower under total anaerobiosis than in environments containing adequate O_2. The rate in water logged soils is intermediate between the two extremes.

When a soil is water logged or flooded there is a shift from aerobic to anaerobic transformation. Formation and accumulation of organic acids *viz.*, acetic, formic, butyric, lactic and succinic acids appear too, these are frequently detrimental to root development. Organic acids accumulate because of the fermentative character of the microflora of wet soils. The anaerobic carbon transformation is thus characterized by the formation of organic acids, alcohols, CH_4 and CO_2 as major end products.

Under anaerobic conditions, decomposition of organic residues takes place by the activity of both mesophilic, thermophilic microorganisms resulting in the production of CO_2, H, ethyl alcohol, and organic acids. Among mesophiles, bacteria are more active than fungi or actinomycetes in cellulolytic activity. They belong to genus Clostridium which are numerous in manure pits but rarely encountered in cultivated arable soils. In compost pits both meso and thermopholic (bacterial and actinomycetes) are important in the breakdown of cellulose substances.

The primary microbial colonizers initially breakdown the complex CH_2O and proteins into organic acids and alcohols. At a later stage, the methane bacteria, which are strict anaerobes, begin to act upon the secondary substrate chiefly lactic and butyric acids and ferment them into CH_4 and CO_2.

Organic matter decomposition serves two important functions

a) Provide energy for growth

b) Supply carbon for the formation of new cell materials

Importance of organic matter decomposition

- Important function is the breakdown of organic matter by which CO_2 available for photosynthesis is replenished
- Any compound that is synthesized biologically is subject to destruction by soil inhabitants, otherwise the compounds would have accumulated in vast amounts on the earth's surface
- Since, organic matter degradation is a property of all heterotrophs, it is commonly used to indicate the level of microbial activity.
- Methods to evaluate the decomposition rate
- Measurement of CO_2 evolution or O_2 uptake
- Determination of decrease in organic matter either chemically or by weight loss
- Observations on disappearance of specific constituents such as cellulose, hemicellulose or lignin.

The end products

The end products of decomposition are CO_2, H_2O, NO_3, SO_4, CH_4, NH_4 and H_2S depending on the availability of oxygen.

Factors influencing the organic matter decomposition

- Organic matter level of the soil
- Cultivation practices
- Temperature
- Moisture
- pH
- Depth
- Aeration
- Nature and abundance of microorganisms involved
- The extent of availability of C, N, P & K and presence of inhibitory substances

C:N ratio

Nitrogen is a key nutrient substance for microbial growth. If the N content of the substrate is high, it is readily utilized and decomposition is faster and if the N is poor, the decomposition is slower, needs additional N. Protein rich substrates are readily decomposed. Low N or wide C : N ratio results slow decay. Optimum level of C: N ratio for maximum decomposition is 20-25 (1.4-1.7% N). Less than this range, more microbial cells, faster mineralization and it likely exceed immobilization. Wider the ratio, lesser microbial cells, slower the immobilization and mineralization increases gradually, resulting in accumulation of ammonia and nitrates. At optimum level, there must be equilibrium between mineralization and immobilization.

Humus

A dark coloured and fairly stable soil organic matter with known and unknown physical and chemical properties is called as humus. It is an integral part of the organic matter complex in soil. Humus can be defined as lingo protein complex containing approximately 45 % lignin compounds, 35 % amino acids, 11 % carbohydrates, 4 % cellulose, 7 % Hemicellulose, 3 % fats, wax, resins and 6 % other miscellaneous substances, including plant growth substances and inhibitors.

Age and composition of the humus are dependent on its origin and environment. Bacterial and algal protoplasm contribute a good deal to the nutritive value of humus. Soil microorganisms take part in humus formation. Some fungi such as Penicillium, Aspergillus and actinomycetes produce dark humus like substances, which serve as structural units for the synthesis of humic substances. The organic fraction of soil is often termed as humus. It is a product of synthetic and decomposing activities of the microflora. Since it contains the organic C and N needed for microbial development, it is the dominant food reservoir.

Humus formation

Once the plant or animal remains fall on are incorporated into the soil, they are subjected to decomposition. From the original residues, a variety of products are formed. As the original materials and the initial products undergo further decomposition, they are converted to brown or black organic complexes. At this stage any trace of the original remains no longer remains. The native organic fraction originates from two sources: the original plant debris entering the soil and the microorganisms with in the soil body. The microorganisms in soil body work upon the former and synthesize microbial protoplasm and new compounds that become part of the organic fraction.

Humus exists in a dynamic state and the chemistry of humus is complex. It has been pointed out that the organic fraction is derived from plant constituents that are modified by the microflora.

Constituents of microbial cells and products of microbial metabolism that is relatively resistant to decay and therefore persist for sometime after death of organism.

The organic fraction contains compounds of C, H, O, N, P, S and small amounts of other elements. Only a small portion of the total is soluble in water, but much can be brought into solution by alkali. Humus contains number of polymerized substances, aromatic, molecules, polysaccharides, ascorbic acids, polymers of uronic acids and P containing compounds. No definite composition can be assigned. It should be considered as a portion of the soil that is composed of a heterogenous group of substances, most having an unknown parentage and an unknown chemical structure. Lignin and lignin derived molecules have long been considered to be of significance in the formation of humus. It is possible either that simple aromatics released in the microbial attack on lignin polymerize to yield constituents of the soil organic fraction or that partially altered lignin itself give rise to humus constituents. The monomeric portions of humus are similar to the constituents of lignin.

Benefits of humic substances

- Improves seed germination, root growth, uptake of minerals by plants and other physiological effects on plant growth
- Increases the enzyme activities involved in plant metabolism. Since humic acid serves as hydrogen acceptors
- Increases the cytochrome oxidase activity in root systems results in growth stimulatory effect (on roots)
- Vigour and yield of plants enhanced
- Humic acid known to influence the grown and proliferation of microorganisms.

Factors Influencing rate of Organic Matter Decomposition

In addition to the composition of organic matter, nature and abundance of microorganisms in soil, the extent of C, N, P and K., moisture content of the soil and its temperature, PH, aeration, C: N ratio of plant residues and presence/absence of inhibitory substances (*e.g.* tannins) *etc.* are some of the major factors which influence the rate of organic matter decomposition.

As soon as plant and animal residues are added to the soil, there is a rapid increase in the activity of microorganisms. These are not true soil organisms, but they continue their activity by taking part in the decomposition of organic matter and thereby release of plant nutrients in the soil. Bacteria are the most abundant organisms playing important role in the decomposition of organic matter. Majority of bacteria involved in decomposition of organic matter are heterotrophs and autotrophs are least in proportions which are not directly involved in organic matter decomposition. Actinomycetes and fungi are also found to play important role in the decomposition of organic matter. Soil algae may contribute a small amount of organic matter through their biomass but they do not have any active role in organic matter decomposition.

a) Aeration: Good aeration is necessary for the proper activity of the microorganisms involved in the decomposition of organic matter. Under anaerobic conditions fungi and actinomycetes are almost suppressed and only a few bacteria (Clostridium) take part in anaerobic decomposition. The rate of decomposition is markedly retarded. It was found that under aerobic conditions 65 percent of the total organic matter decomposes during six months, while under anaerobic conditions only 47 percent organic matter can be decomposed during the same period. Anaerobic decomposition of organic matter results into the production of large quantity of organic acids and evolution of gases like methane (CH_4) hydrogen (H_2) and carbon dioxide (CO_2).

b) Temperature: The rate of decomposition is more rapid in the temperature range of 30° to 40°. At temperatures below or above this range, the rate of decomposition is markedly retarded. Appreciable organic mater decomposition occurs at 25° C and further fluctuation in the soil temperature has little effect on decomposition.

c) Moisture: Adequate soil moisture *i.e.* about 60 to 80 percent of the water-holding capacity of the soil is must for the proper decomposition of organic matter. Too much moisture leads to insufficient aeration which results in the reduced activity of microorganisms and there by checks the rate of decomposition.

d) Soil PH/soil reaction: Soil PH affects directly the kind, density and the activity of fungi, bacteria & actinomycetes involved in the process of decomposition and thereby rate of decomposition of organic matter. The rate of decomposition is more in neutral soils than that of acidic soils. Therefore, treatment of acid soils with lime can accelerate the rate of organic matter decomposition.

e) C: N ratio: C: N ration of organic matter has great influence on the rate of decomposition. Organic matter from diverse plant-tissues varies widely in their C: N ratio (app. 8-10 %). The optimum C: N ratio in the range of 20-25 is ideal for maximum decomposition, since a favorable soil environment is created to bring about equilibrium between mineralization and immobilization processes. Thus, a low nitrogen content or wide C'.N ratio results into the slow decomposition. Protein rich, young and succulent plant tissues are decomposed more rapidly than die protein-poor, mature and hard plant tissues. Therefore, C:N ratio of organic matter as well as soil should be narrow/less for better and rapid decomposition. Thus, high aeration, mesophilic temperature range, optimum moisture, neutral/alkaline soil reaction and narrow C: N ratio of soil and organic matter are required for rapid and better decomposition of organic matter.

Microbiology of decomposition of various constituents in organic matter

When plant and animal residues are added to the soil, the various constituents of the soil organic matter are decomposed simultaneously by the activity of microorganisms and carbon is released as CO_2, and nitrogen as NH_4—>NO_3 for the use by plants. Other nutrients are also converted into plant usable forms. This process of release of nutrients from organic matter is called mineralization. The insoluble plant residues constitute the part of humus and soil organic matter complex. The final product of aerobic decomposition is CO_2 and that of anaerobic decomposition are Hydrogen, Ethyl Alcohol (CH_4), various organic acids and carbon dioxide (CO_2). Soil organisms use organic matter as a source of energy and food.

The process of decomposition is initially fast, but slows down considerably as the supply of readily decomposable organic matter gets exhausted. Sugars, water-soluble nitrogenous compounds, amino acids, lipids, starches and some of the hemicellulases are decomposed first at rapid rate, while insoluble compounds such as cellulose, hemicellulose, lignin, proteins *etc.* which forms the major portion of organic matter are decomposed later slowly. Thus, the organic matter added to the soil is converted by oxidative decomposition to simpler substances which are made available in stages for plant growth and the residue is transformed into humus.

The microbiology of decomposition/degradation of some of the major constituents (*viz.* Cellulose, Hemicellulose, Lignin, Proteins *etc.*) of soil organic matter/plant residues are discussed in brief in the following paragraphs.

a) Decomposition of Cellulose

Cellulose is the most abundant carbohydrate present in plant residues/ organic matter in nature. When cellulose is associated with pentosans (*e.g.* xylans & mannans) it undergoes rapid decomposition, but when associated with lignin, the rate of decomposition is very slow. The decomposition of cellulose occurs in two stages: (i) in the first stage the long chain of cellulase is broken down into cellobiase and then into glucose by the process of hydrolysis in the presence of enzymes cellulase and cellobiase, and (ii) in second stage glucose is oxidized and converted CO_2 and water.

$$1.\ \text{Cellulose} \xrightarrow[\text{Hydrolysis}]{\text{Cellulase}} \text{Cellobiose} \xrightarrow[\text{Hydrolysis}]{\text{Cellobiase}} \text{Glucose}$$

$$2.\ \text{Glucose} \xrightarrow{\text{Oxidation}} \text{Organic Acids} \xrightarrow{\text{Oxidation}} CO_2 + H_2O$$

The intermediate products formed/released during enzymatic hydrolysis of cellulose (*e.g.* cellobiose and glucose) are utilized by the cellulose-decomposing organisms or by other organisms as source of energy for biosynthetic processes. The cellulolytic microorganisms responsible for degradation of cellulose through the excretion of enzymes (cellulase & Cellobiase) are fungi, bacteria and actinomycetes.

It is a CH_2O composed of glucose units bound by β-linkage at carbon 1 and 4 of the sugar molecule. The cellulose concentration of higher plants is never fixed. It is a polymer of glucose and abundant organic material in nature changes with age and type of plants. Woody materials have more cellulose and succulent tissues have poor, but the concentration increases as the plant matures. Cellulose breakdown in soil is influenced by several environmental factors.

Aerobic organisms convert cellulose in to 2 major products: CO_2^+ cell substance, but certain group release small amount of organic acids. It is however resistant to microbial decomposition. When cellulose is associated with pentosans (xylan, mannans) it undergoes rapid decomposition. When associated with lignin, the decomposition rate is very low. Degradation is by the enzymes that convert cellulose into glucose.

(Exoenzymes)

Exoglucanase

Endoglucanase

- glucosidase (cellulase complex)

$$\text{Native cellulose} \xrightarrow[\text{Endoglucanase}]{\text{Exoglucanase}} \text{Amorphos cellulose + cellobiose}$$

$$\downarrow \text{Endogluconase}$$

β-glucosidases

D- Glucose ← Cellobiose

(cellobiase)

Most cellulolytic bacteria do not excrete significant amounts of cellulase but fungi are found to excrete these enzymes. The soluble sugars released by enzymatic hydrolysis are later utilized by the same or other microorganisms for biosynthetic purpose.

b) Decomposition of Hemicelluloses

Polymer of simple sugars such as pentose, hexose and uronic acids. It may be either homo or hetero polymers. When they are added to soil, degradation takes place at faster rate in initial stages. The hemicelluloses such as mannans are decomposed rapidly while galactons (polymer of galactose) are decomposed slowly. Many soil microorganisms utilize hemicellulose in both aerobic as well as anaerobic conditions. The microbial degradation occurs through the agency of extracellular enzymes called hemicellulases. Hemicelluloses are water-soluble polysaccharides and consists of hexoses, pentoses, and uronic acids and are the major plant constituents second only in quantity of cellulose, and sources of energy and nutrients for soil microflora.

When subjected to microbial decomposition, hemicelluloses degrade initially at faster rate and are first hydrolyzed to their component sugars and uronic acids. The hydrolysis is brought about by number of hemicellulolytic enzymes known as "hemicellulases" excreted by the microorganisms. On hydrolysis hemicelluloses are converted into soluble monosaccharide/sugars (*e.g.* xylose, arabinose, galactose and mannose) which are further convened to organic acids, alcohols, CO2 and H2O and uronic acids are broken down to pentoses and CO2. Various microorganisms including fungi, bacteria and actinomycetes both aerobic and anaerobic are involved in the decomposition of hemicelluloses.

c) Lignin Decomposition

Lignin is the third most abundant constituent of plant tissues, and accounts about 10-30 percent of the dry matter of mature plant materials. Lignin content of young plants is low and gradually increases as the plant grows old. It is one of the most resistant organic substances for the microorganisms to degrade however certain Basidiomycetous fungi are known to degrade lignin at slow rates. Complete oxidation of lignin result in the formation of aromatic compounds such as syringaldehydes, vanillin and ferulic acid. The final cleavages of these aromatic compounds yield organic acids, carbon dioxide, methane and water.

It consists of heterocyclic aromatic organic molecules containing C, H and O. The degradation is very slow and rate of decomposition depends on the presence of other compounds such as cellulose and hemicellulose acid. Lignin is highly resistant to microbial degradation. Degradation is a complex process.

Lignin → coniferyl ether → coniferyl alcohol → coniferyl aldehyde → vanillin

→ vannillic acid → protocatechuic acid → ring cleavage

d) Protein Decomposition

Proteins are complex organic substances containing nitrogen, sulphur, and sometimes phosphorus in addition to carbon, hydrogen and oxygen. During the course of decomposition of organic matter, proteins are first hydrolyzed to a number of intermediate products *e.g.* Proteases, peptides *etc.* collectively known as polypeptides

The intermediate products so formed are then hydrolyzed and broken down ultimately to individual amino acids, or ammonia and amides. The process of hydrolysis of proteins to amino acids is known as "aminization or ammonification", which is brought about by certain enzymes, collectively known as "proteases" or "proteolytic" enzymes secreted by various microorganisms. Amino acids and amines are further decomposed and converted into ammonia. During the course of ammonification, various organic acids, alcohols, aldehydes *etc.* are produced which are further decomposed finally to produce carbon dioxide and water.

All types of microorganisms, bacteria, fungi, and actinomycetes are able to bring about decomposition of proteins. In acid soils, fungi are pre-dominant, while in neutral and alkaline soils bacteria are dominant decomposers of proteins.

Phosphorus Cycle or Transformation

Phosphorus is only second to nitrogen as an inorganic nutrient required by both plants and microorganisms. It stimulates growth of young plants, giving them a good and vigorous start. Phosphorus management and nutrition has both economic and environmental implications. Phosphate constitutes nearly 0.1% of the earth's crust. They occur in soil in inorganic and organic forms. The organic phosphorus containing compounds are derived from plants and microorganisms and are composed of nucleic acids, phospholipids, lecithin, phytin and related compounds. In soil, 15-85 % of the total P is organic and soils rich in organic matter contain abundant organic P. The inorganic forms are derived from parent rocks or through fertilizers application and manuring with bone meal. They are soluble in water when present as phosphates of Na, K, Ca, Mg *etc.* Phosphorus exists mainly as apatites, with the basic formula $M_{10}(PO_4)_6X_2$. Commonly the mineral (M) is Ca, less often Al or Fe. The anion (X) is either F^-, Cl^-, OH^- or CO^{2-}_3. Diverse combinations of M and X results in 200 forms of P.

It is a major constituent of nucleic acids in all living systems essential in the accumulation and release of energy during cellular metabolism. This element is added to the soil in the form of chemical fertilizers, or in the form of organic phosphates present in plant and animal residues. In cultivated soils it is present in abundance (*i.e.* 1100 kg/ha), but most of which is not available to plants, only 15 % of total soil phosphorus is in available form. Both inorganic and organic phosphates exist in soil and occupy a critical position both in plant growth and in the biology of soil.

Microorganisms are known to bring a numberof transformations of phosphorus,these include:

i. Altering the solubility of inorganic compounds of phosphorus,

ii. Mineralization of organic phosphate compounds into inorganic phosphates,

iii. Conversion of inorganic, available anion into cell components *i.e.* an immobilization process

iv. Oxidation or reduction of inorganic phosphorus compounds. Of these mineralization and immobilization are the most important reactions / processes in phosphorus cycle.

Insoluble inorganic compounds of phosphorus are unavailable to plants, but many microorganisms can bring the phosphate into solution. Soil phosphates are rendered available either by plant roots or by soil microorganisms through secretion of organic acids (*e.g.* lactic, acetic, formic, fumaric, succinic acids *etc*). Thus, phosphate-dissolving / solubilizing soil microorganisms (*e.g.* species of *Pseudomonas, Bacillus, Micrococcus, Mycobacterium, Flavobacterium, Penicillium, Aspergillus, Fusarium* etc.) plays important role in correcting phosphorus deficiency of crop plants. They may also release soluble inorganic phosphate (H2PO4), into soil through decomposition of phosphate-rich organic compounds.

Solubilization of phosphate by plant roots and soil microorganisms is substantially influenced by various soil factors, such as PH, moisture and aeration.

In neutral or alkaline soils solubilization of phosphate is more as compared to acidic soils. Many phosphates solubilizing microorganisms are found in close proximity of root surfaces and may appreciably enhance phosphate assimilation by higher plants.

By their action, fungi bacteria and actinomycetes make available the organically bound phosphorus in soil and organic matter and the process is known as mineralization. On the other hand, certain microorganisms especially bacteria assimilate soluble phosphate and use for cell synthesis and on the death of bacteria, the phosphate is made available to plants. A fraction of phosphate is also lost in soil due to leaching. One of the ways to correct deficiency of phosphorus in plants is to inoculate seed or soil with commercial preparations (*e.g.* Phosphobacterin) containing phosphate - solubilizing microorganisms along with phosphatic fertilizers.

Mineralization of phosphate is generally rapid and more in virgin soils than cultivated land. Mineralization is favored by high temperatures (thermophilic range) and more in acidic to neutral soils with high organic phosphorus content. The enzyme involved in mineralization (cleavage) of phosphate from organic phosphorus compound is collectively called as **"Phospatases"**.

The commercially used species of phosphate solubilizing bacteria and fungi are: *Bacillus polymyxa, Bacillus megatherium. Pseudomonas strita, Aspergillus, Penicllium avamori and Mycorrhiza*

Sources of Phosphate

1) Greatest reservoirs of phosphate are the insoluble ferric and calcium phosphates

in rocks (rock phosphate) in combination with calcium, iron and aluminium. Slow process of weathering releases phosphates to the soil.

2) Phosphates are also added to the soil by man in the form of artificial fertilizers.

3) Some quantity of phosphorus is deposited in deep sediments in oceans and is brought to the cycle by upwelling (but not extensive).

- ❑ Microorganisms bring about a number of transformations of this element.
- ❑ Altering the solubility of inorganic compounds of P
- ❑ Mineralization of organic compounds with the release of inorganic phosphate
- ❑ Converting the inorganic, available anion into cell components, an immobilization process (analogous to that occurring with N)
- ❑ Bringing about an oxidation or reduction of inorganic P compoun

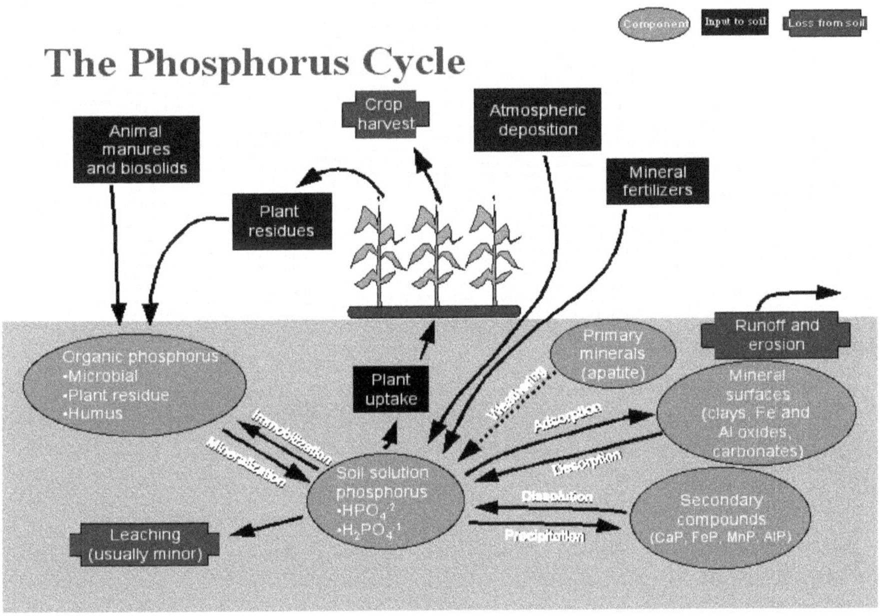

1. Solubilization of inorganic phosphorus

In cultivated soil, P is present abundantly about 1100 kg/ha, but most of them as not available to plants; only about 1% of the total P is in available form. Inorganic phosphorus is negatively charged in most soils. Because of its particular chemistry, phosphorus reacts readily with positively charged iron (Fe), aluminum (Al), and calcium (Ca) ions to form relatively insoluble substances. When this occurs, the phosphorus is considered fixed or tied up. The solubility of the various inorganic phosphorus compounds directly affects the availability of phosphorus for plant growth. The solubility is influenced by the soil pH. Soil phosphorus is most available for plant use at pH values of 6 to 7. When pH is less than 6, plant available phosphorus becomes increasingly tied up in aluminum phosphates. As soils become more acidic (pH below 5), phosphorus is fixed in iron phosphates.

Insoluble inorganic compounds of P are largely unavailable to plants, but many microorganisms can bring the PO_4 into solution. Pseudomonas, Microoccus,Bacillus, Fusarium, Aspergillus solubilize calcium salts, iron, aluminum, magnesium manganese phosphate. P is solubilized by the production of organic acids. The acids convert $Ca_3 (PO_4)_2$ to di and monobasic phosphates and releases P to plants. These organisms secrete high amount of phosphatase enzymes which also solubilize bound form of phosphates.

2. Mineralization of organic phosphorus

Organic form of P is the larger reservoir of P in soil. By the action of bacteria, fungi and actinomycetes, bound element in remains of the vegetation and in soil organic matter is made available to succeeding generations of plants. Among the organic phosphorus compounds, lecithin, nucleic acids and phytin occupy prominent place. Lecithin contains 9.39 % P_2O_5, 1.6% N and 65.36% C.

It is a process of conversion of organic forms of phosphorus into inorganic available forms of P. Mineralization is mediated by the enzymes called phosphatases. These enzymes cleave phosphorus from more frequently encountered organic substrates. Phytases liberate PO_4 from phytic acid or its Ca-Mg salt, Phytin. They remove PO_4-s, one at a time, yield penta, tetra, di and mono PO_4 and then finally free inositol. Bacillus, Pseudomonas, Aspergillus, Penicillium, Rhizopuscan synthesize this enzyme. Mycorrhizal (fungi) are also able to mineralize the organic forms of P and increases P uptake by the plants.

3. Immobilization

Process of assimilation of P into microbial nucleic acids, phospholipids or other protoplasmic substances is called immobilization. It leads to the accumulation of non-utilizable forms of the element.

4. Oxidation reduction reactions

Oxidative process Biological oxidation of reduced phosphorus compounds into oxidized state.

Phosphite ($HPO_3^=$) is oxidized to phosphate. A number of hetertrophic (bacteria), (fungi) & (actinomycetes) utilize phosphite as sole P source. Hypophosphites ($HPO_2^=$) can also be oxidized to phosphate by heterotrophs.

$$HPO_3^= \rightarrow HPO_4^=$$

$$HPO_2^= \rightarrow HPO_4^=$$

Reductive process, reductive pathway has also been functioned. PO_4 is reduced to phosphite and hypophosphite.

$$H_3PO_4 \rightarrow H_3PO_3 \rightarrow H_3PO_2$$

Clostridium.butyricum, E. coli form phosphite and hypophosphite from orthophosphate. It is biochemically analogue to the process of denitirification. Only little information is available about this process.

P exist in an organic form in the protoplasm on the death of living organism, this (P) is changed to inorganic phosphoric acid. This is soon converted into insoluble salts of Ca, Fe, Mg and Al. Phosphorus thus alternates between organic and inorganic, and soluble and insoluble forms. In soluble P is solubilized by various acids produced by micro organism.

Microbial activities involved in the cycling of C, N and P are absolutely essential for maintenance of soil fertility.

Sulphur cycle

Microbial transformations of sulfur (S) are even more complex than those of nitrogen because of the large number of oxidation states of S and the fact that several transformations of S also occur abiotically. Although a number of oxidation states of S are possible, only three are significant in nature, 2 (sulfhydryl, R–SH, and sulfide, HS-), 0 (elemental sulfur, S0), and 6 (sulfate, SO_4 2-). The bulk of S Earth's S is in sediments and rocks in the form of sulfate minerals, primarily gypsum ($CaSO_4$) and sulfide minerals (pyrite, FeS2), but the oceans constitute the most significant reservoir of $SO_4$2- in the biosphere. A significant amount of S, in particular sulfur dioxide (SO_2, a gas), enters the S cycle from human activities, primarily the burning of fossil fuels.

Hydrogen Sulfide and Sulfate Reduction

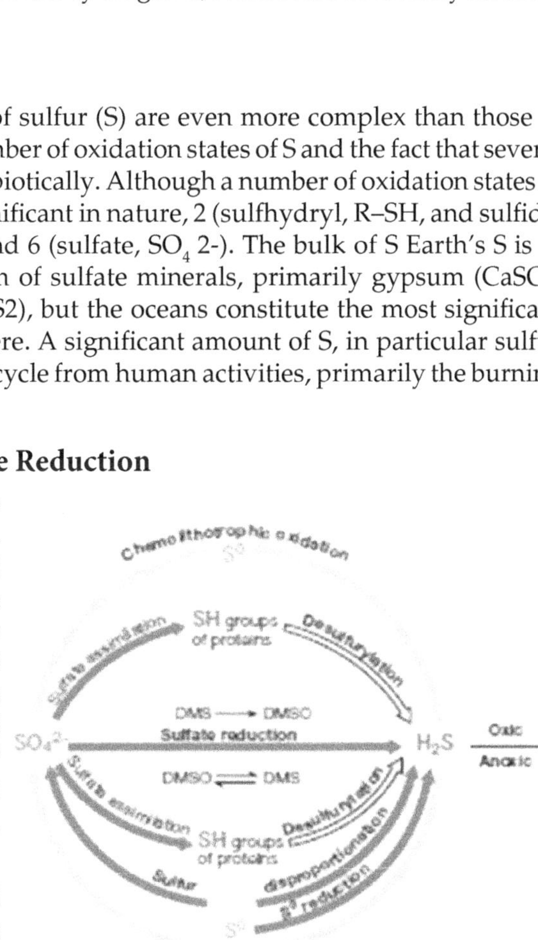

A major volatile S gas is hydrogen sulfide (H_2S). Hydrogen sulfide is produced from bacterial sulfate reduction or is emitted from sulfide springs and volcanoes. Although H_2S is volatile, different forms exist depending on pH: H_2S predominates below pH 7 and the nonvolatile HS- and S2- predominate above pH 7. Collectively,

H_2S, HS, and S_2 are referred to as "sulfide." Sulfate-reducing bacteria are a large and highly diverse group and are widespread in nature. However, in anoxic habitats such as freshwater sediments and many soils, sulfate reduction is SO_4 2--limited. Moreover, because organic electron donors (or H_2, which is a product of the fermentation of organic compounds) are needed to support sulfate reduction, it only occurs where significant amounts of organic material are present.

In marine sediments, the rate of $SO_4$2- reduction is typically C limited and can be greatly increased by an influx of organic matter. This is important because the disposal of sewage or garbage in the oceans or coastal regions can trigger sulfate reduction. Hydrogen sulfide is toxic to many plants and animals and therefore its formation is potentially detrimental (sulfide is toxic because it combines with the iron of cytochromes and blocks respiration). Sulfide is commonly detoxified in nature by combination with iron, forming the insoluble minerals FeS (pyrrhotite) and FeS_2 (pyrite). The black color of sulfidic sediments or sulfate-reducing bacterial cultures is due to these metal sulfide minerals.

Sulfide and Elemental Sulfur

Oxidation–Reduction

Under oxic conditions, sulfide rapidly oxidizes spontaneously at neutral pH. Sulfur-oxidizing chemolithotrophic bacteria, most of which are aerobes, can catalyze the oxidation of sulfide. However, because of the rather rapid spontaneous reaction, microbial sulfide oxidation is significant only in areas where H_2S emerging from anoxic environments meets air. Where light is available, there can be anoxic oxidation of sulfide, catalyzed by the phototrophic purple and green sulfur bacteria

Elemental sulfur is chemically stable but is readily oxidized by sulfur-oxidizing chemolithotrophic bacteria such as Thiobacillus and Acidithiobacillus. Because S0 is insoluble, the bacteria that oxidize it must attach to the S0 crystals to obtain their substrate. The oxidation of S0 forms sulfuric acid (H_2SO_4), and thus S0 oxidation characteristically lowers the pH in the environment, sometimes drastically. For this reason, So is sometimes added to alkaline soils as an inexpensive and natural way to lower the pH, relying on the ubiquitous sulfur chemolithotrophs to carry out the acidification process. Elemental sulfur can be reduced as well as oxidized. The reduction of S0 to sulfide (a form of anaerobic respiration) is a major ecological process, especially among hyperthermophilic Archaea. Although sulfate-reducing bacteria can also reduce S0, in sulfidic habitats most S0 is reduced by the physiologically specialized sulfur reducers, organisms that are incapable of $SO_4$2- reduction. The habitats of the sulfur reducers are generally those of the sulfate reducers, so from an ecological standpoint, the two groups form a metabolic guild unified by their formation of H_2S.

Organic Sulfur Compounds

In addition to inorganic forms of S, several organic S compounds are also cycled in nature. Many of these foul-smelling compounds are highly volatile and can thus enter the atmosphere.

The most abundant organic S compound in nature is dimethyl sulfide (CH_3—S—CH_3); it is produced primarily in marine environments as a degradation product of dimethylsulfoniopropionate, a major osmoregulatory solute in marine algae. This compound can be used as a carbon source and electron donor by microorganisms and is catabolized to dimethyl sulfide and acrylate. The latter, a derivative of the fatty acid propionate, is used to support growth.

Dimethyl sulfide released to the atmosphere undergoes photochemical oxidation to methanesulfonate (CH_3SO_3), SO_2, and $SO_4$2-. By contrast, CH_3—S—CH_3 produced in anoxic habitats can be microbially transformed in at least three ways: (1) by methanogenesis (yielding CH_4 and H_2S), (2) as an electron donor for photosynthetic CO_2 fixation in phototrophic purple bacteria (yielding dimethyl sulfoxide, DMSO), and (3) as an electron donor in energy metabolism in certain chemoorganotrophs and chemolithotrophs (also yielding DMSO). DMSO can be an electron acceptor for anaerobic respiration, producing CH_3—S—CH_3. Many other organic S compounds affect the global sulfur cycle, including methanethiol (CH_3SH), dimethyl disulfide (H_3C—S—S—CH_3), and carbon disulfide (CS_2), but on a global basis, CH_3—S—CH_3 is the most significant.

Sulphur Transformations

Sulphur is the most abundant and widely distributed element in the nature and found both in free as well as combined states. Sulphur, like nitrogen is an essential element for all living systems. In the soil, sulphur is in the organic form (sulphur containing amino acids-cystine, methionine, proteins, polypeptides, biotin, thiamine *etc*) which is metabolized by soil microorganisms to make it available in an inorganic form (sulphur, sulphates, sulphite, thiosulphale, *etc*) for plant nutrition. Of the total sulphur present is soil only 10-15% is in the inorganic form (sulphate) and about 75-90 % is in organic form.

Cycling of sulphur is similar to that of nitrogen. Transformation / cycling of sulphur between organic and elemental states and between oxidized and reduced states is brought about by various microorganisms, specially bacteria- Thus "the conversion of organically bound sulphur to the inorganic state by microorganisms is termed as mineralization of sulphur". The sulphur / sulphate, thus released are either absorbed by the plants or escapes to the atmosphere in the form of oxides.

Various transformations of the sulphur in soil results mainly due to microbial activity, although some chemical transformations are also possible (*e.g.* oxidation of iron sulphide) the major types of transformations involved in the cycling of sulphur are:

1. Mineralization
2. Immobilization
3. Oxidation
4. Reduction

1. Mineralization: The breakdown / decomposition of large organic sulphur compounds to smaller units and their conversion into inorganic compounds (sulphates) by the microorganisms. The rate of sulphur mineralization is about 1.0 to 10.0 percent / year.

2. Immobilization: Microbial conversion of inorganic sulphur compounds to organic sulphur compounds.

3. Oxidation: Oxidation of elemental sulphur and inorganic sulphur compounds (such as h2S, sulphite and thiosulphale) to sulphate (SO4) is brought about by chemoautotrophic and photosynthetic bacteria.

When plant and animal proteins are degraded, the sulphur is released from the amino acids and accumulates in the soil which is then oxidized to sulphates in the presence of oxygen and under anaerobic condition (water logged soils) organic sulphur is decomposed to produce hydrogen sulphide (H2S). H2S can also accumulate during the reduction of sulphates under anaerobic conditions which can be further oxidized to sulphates under aerobic condition.

a) $2S + 3O_2 + 2H_2O \longrightarrow 2H_2SO_4 \longrightarrow 2H(+) + SO_4$ (Aerobic)

b) $CO_2 + 2H_2S \longrightarrow (CH_2O) + H_2O + 2S$

OR $H_2 + S + 2\,CO_2 + H_2O \longrightarrow H_2SO_4 + 2(CH_2O)$ (anaerobic)

The members of genus Thiobacillus (obligate chemolithotrophic, non photosynthetic) eg, T. ferrooxidans and T. thiooxidans are the main organisms involved in the oxidation of elemental sulphur to sulphates. These are aerobic, non-filamentous, chemosynthetic autotrophs. Other than Thiobacillus, heterotrophic bacteria (Bacillus, Pseudomonas, and Arthrobacter) and fungi (Aspergillus, Penicillium) and some actinomycetes are also reported to oxidize sulphur compounds. Green and purple bacteria (Photolithotrophs) of genera Chlorbium, Chromatium. Rhodopseudomonas are also reported to oxidize sulphur in aquatic environment.

Sulphuric acid produced during oxidation of sulphur and H: S is of great significance in reducing the PH of alkaline soils and in controlling potato scab and rot diseases caused by Streptomyces bacteria. The formation of sulphate / Sulphuric acid is beneficial in agriculture in different ways : i) as it is the anion of strong mineral acid (H_2SO_4) can render alkali soils fit for cultivation by correcting soil PH. ii) solubilize inorganic salts containing plant nutrients and thereby increase the level of soluble phosphate, potassium, calcium, magnesium *etc.* for plant nutrition.

4. Reduction of Sulphate: Sulphate in the soil is assimilated by plants and microorganisms and incorporated into proteins. This is known as "assimilatory sulphate reduction". Sulphate can be reduced to hydrogen sulphide (H2S) by sulphate reducing bacteria (eg. Desulfovibrio and Desulfatomaculum) and may diminish the availability of sulphur for plant nutrition. This is "dissimilatory sulphate reduction" which is not at all desirable from soil fertility and agricultural productivity view point.

Dissimilatory sulphate-reduction is favored by the alkaline and anaerobic conditions of soil and sulphates are reduced to hydrogen sulphide. For example, calcium sulphate is attacked under anaerobic condition by the members of the genus Desulfovibrio and Desulfatomaculum to release H2 S.

$CaSO_4 + 4H_2 \longrightarrow Ca\,(OH)_2 + H_2S + H_2O.$

Hydrogen sulphide produced by the reduction of sulphate and sulphur containing amino acids decomposition is further oxidized by some species of green and purple phototrophic bacteria (*e.g.* Chlorobium, Chromatium) to release elemental sulphur.

$$CO_2 + 2H_2 + H_2S \xrightarrow[\text{Enzyme}]{\text{Light}} (CH_2O) + H_2O + 2S.$$

Carbohydrate Sulphur

The predominant sulphate-reducing bacterial genera in soil are Desulfovibrio, Desulfatomaculum and Desulfomonas. (All obligate anaerobes). Amongst these species Desulfovibrio desulfuricans are most ubiquitous, non-spore forming, obligate anaerobes that reduce sulphates at rapid rate in waterlogged / flooded soils. While species of Desulfatomaculum are spore forming, thermophilic obligate anaerobes that reduce sulphates in dry land soils. All sulphate-reducing bacteria excrete an enzyme called **"desulfurases" or "bisulphate Reductase"**. Rate of sulphate reduction in nature is enhanced by increasing water levels (flooding), high organic matter content and increased temperature.

Iron Cycle or Transformation

Iron exists in nature either as ferrous (Fe++) or ferric (Fe+++) ions. Ferrous iron is oxidized spontaneously to ferric state, forming highly insoluble ferric hydroxide. Plants as well as microorganisms require traces of iron, manganese copper, zinc, molybdenum, calcium boron, cobalt *etc*. Iron is always abundant in terrestrial habitats, and it is oftenly in an unavailable form for utilization by plants and leads to the serious deficiency in] plants.

Soil microorganisms play important role in the transformations of iron in al number of distinctly different ways such as:

A. Certain bacteria oxidize ferrous iron to ferric state which precipitate as ferric hydroxide around cells

B. Many heterotrophic species attack on in soluble organic iron salts and convert into inorganic salts

C. Oxidation-reduction potential decreases with microbial growth and which leads to the formation of more soluble ferrous iron from highly insoluble ferric ions

D. Number of bacteria and fungi produce acids such as carbonic, nitric, Sulphuric and organic acids which brings iron into solution

E. Under anaerobic conditions, the sulfides formed from sulphate and organic sulphur compounds remove the iron from solution as ferrous sulfide

F. As microbes liberate organic acids and other carbonaceous products of metabolism which results in the formation of soluble organic iron complex. Thus, iron may be precipitated in nature and immobilized by iron oxidizing bacteria under alkaline soil reaction and on the other hand solubilization of iron may occur through acid] formation.

Some bacteria are capable of reducing ferric iron to ferrous which lowers the oxidation-reduction potential of the environment (*e.g*. Bacillus, Clostridium, Klebsiella *etc*). However, some chemoautotrophic iron and sulphur bacteria such as Thiobacillus ferroxldans and Ferrobacitlus ferrooxidans can oxidize ferrous iron to ferric hydroxide which accumulates around the cells.

Most of the aerobic microorganisms live in an environment where iron exists in the oxidized, insoluble ferric hydroxide form. They produce iron-binding compounds in order to take up ferric iron. The iron-binding or chelating compounds / ligands produced by microorganisms are called **"Siderophores"**. Bacterial siderophores may act as virulence factors in pathogenic bacteria and thus, bacteria that secrete siderophores are more virulent than non- siderophores producers. Therefore, siderophore producing bacteria can be used as biocontrol agents *e.g.* Fluorescent pseudomonads used to control Pythium, causing damping-off diseases in seedlings. Recently Vascular - Arbusecular – Mycorrhiza (VAM) has been reported to increase uptake of iron.

The ironcycle (Fe) is the biogeochemical cycle of iron through landforms, the atmosphere, and oceans. The iron cycle affects dust deposition and aerosol iron bioavailability.

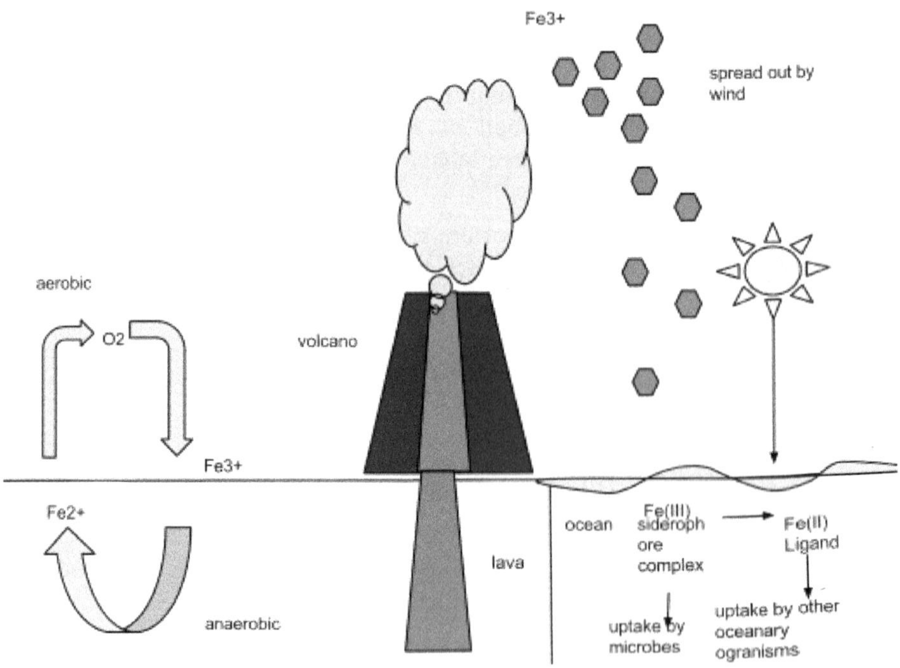

Unit II: Environmental Microbiology

9

Introduction to Environmental Microbiology

Environmental Microbiology

1. Microbes are nature's decomposers. The variety of metabolic abilities in microbes is enormous, and includes microbes that can degrade or mitigate all sorts of human products and activities, from oil spills to pesticide runoff to toxic waste.
2. Environmental Microbiology seeks to find ways to maximize the efficiency of microbes in helping to remove various kinds of wastes (*e.g.* sewage treatment), or to minimize the opportunities of microbes to produce problems (*e.g.* water treatment).
3. Environmental Microbiology is a growing field, often brings together issues of concern to engineers, geologists and hydrologists, microbiologists, and public health officials.

Sewage Treatment

1. Until 1900's, human wastes were simply dumped as raw sewage into the nearest outhouse, stream, or river. As connection of sewage to diseases such as cholera became clear, public policy changed to required water treatment. This had major impact on reduction of many diseases.
2. Sewage = mix of domestic + industrial waste plus drainage water from rainfall. Contains many microbes, mostly harmless but some pathogens from humans or animals.Can include bacteria such as Vibrio cholera, Shigella dysenteriae, enteropathogenic strains of E. coli and B. cereus, viruses such as Hepatitis A, many more.
3. Sewage treatment: goal is to get rid of pathogens, also reduce organic content of effluent to a low level.
4. Primary wastewater treatment: use screens to remove large objects (plastic bags, wads of paper, *etc.*), then move water to large tank to allow settling of heavier particulate matter as sludge.

5. Secondary wastewater treatment: modern facilities use "activated sludge process". After moving water from primary settling tank, bubble air through a secondary tank. Aerobic microbes grow and break down organic matter in the tank. Then move water to another tank called the secondary clarifier, where solids settle and are added to sludge. Clarified liquid is treated with chlorine to kill remaining microbes, then discharged as clear liquid into nearest river. Anaerobic bacteria break down organic matter, produce lots of fermentation products. Methanogens grow on these wastes and produce methane gas as waste. This can be trapped and used as fuel (useful in developing countries).
6. Tertiary wastewater treatment: Secondary wastewater treatment does not remove inorganic ions, such as NH3, PO4-3, SO4-2. Wastewater can enrich local waters to create eutrophic conditions, including algal blooms and sufficient loss of oxygen that fish die. To prevent this, a few water treatment systems use additional steps to remove ammonia and phosphate, using additional processing tanks in which specific bacteria are used to remove ammonia and phosphate.

Bioremediation

1. Expanded use of chemicals in industry has produced major new problems of environmental pollution. U.S. alone has over 50,000 hazardous waste sites. Entire communities have been evacuated because of accumulated toxic wastes. Groundwaters are often polluted as well, including toxic chemicals such as commercial solvents used to degrease machinery or in "dry cleaning". Fertilizers and pesticides are often found in water downstream from agriculture.
2. Bioremediation = use of living organisms to promote destruction of environmental pollutants. See example of anaerobic toluene degrader .
3. Typically, native microbes are used (rather than introduced or genetically mofified organisms). Rather than waiting for "nature to take its course", try to speed up the process. How?
4. "Pump and treat". One way to speed bioremediation. Pump groundwater to surface, add nutrients (*e.g.*, O_2, methane), reinject into contaminated zone. In some cases, can inject
5. Bioreactor. Another technique. Put contaminated soil or groundwater into an industrial-sized fermentor, add appropriate microbes to degrade materials, keep adding more substrate over time. This works well for very toxic chemicals such as chlorinated compounds (PCBs).
6. Microbiology of water

Drinking water is obtained either from surface sources such as rivers, lakes or from underground water. Such natural waters are likely to be polluted with domestic and industrial wastes. Although water purification systems envisage protection from pollution, sometimes, the water supply can become a potential carrier of pathogenic organisms and endanger public health. A number of diseases such as cholera, typhoid, viral hepatitis *etc.*, are known to be water borne. These pathogens are commonly transmitted through drinking water and cause infection of the intestinal tract. It is therefore, necessary to employ treatment facilities to purify water and to provide safe drinking water (Potable water). Water that is free from diseases producing organism and chemical substances deleterious to health is called potable water The main operations employed in water purification to produce potable water are: (i) sedimentation, (ii) filtration, and (iii) chlorination. Sedimentation removes large particulate matter which settles at the bottomMost microorganisms are removed during coagulation with aluminium sulphate and sand filtration and subsequent treatment of water with chlorine (0.2 – 2 mg free chlorine per liter) will ensure its potability.

Main Operation in drinking water purification

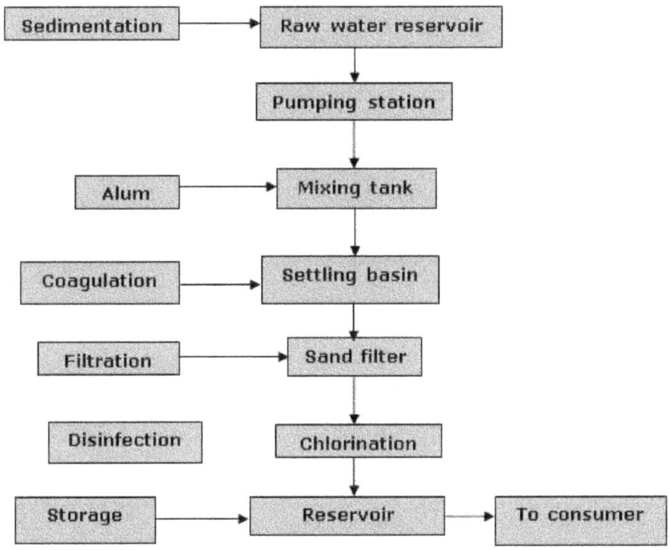

Microbiological Quality: Water can be perfectly clear in appearance and free from odour and taste and yet, be contaminated by microorganisms. Pathogenic organisms enter into water through sewage contamination or discharges from animals or humans into the reservoirs. The coliforms (E.coli and related organisms), Streptococcus faecalis and Clostridium perfringens which are normal inhabitants of the large intestine of animals and humans enter water supplies through faecal contamination. The presence of any of these bacterial species in water is evidence of sewage or faecal pollution. Techniques are available by which the presence of these specific groups can be easily identified. The routine bacteriological examination contains consists of (i) Plate count to determine the number of bacteria present, and (ii) biochemical test to reveal the presence of coliform bacteria since these are indicator organisms for fecal contamination. A variety of other bacteria and organisms which may not be serious pathogens including faecal streptococci, slime forming bacteria, Sulphur bacteria, algae *etc.* may also cause problems of odour, color and taste and it is essential that be eliminated from the drinking water.

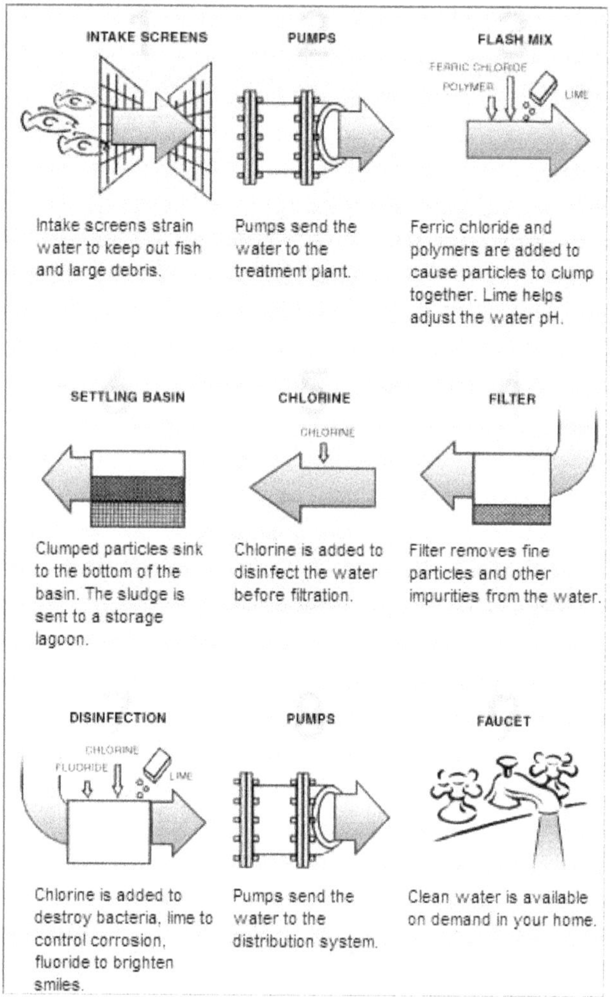

Laboratory tests for detecting contamination by coliforms presumptive test

Bioconversion of Agricultural Wastes

Bioconvertion refers to the recycling and reprocessing of organic waste materials using the system of biological decomposition or biodegradation into value added products.

Composting is one way of convertion of organic waste into manures. It is defined as a method of solid waste management whereby the organic component of the solid waste stream is biologically decomposed under controlled conditions to a state in which it can be handled, stored or applied to the land without adversely affecting the environment. It is predominantly aerobic process, that allow the development of thermophilic temperatures as a result of biologically produced heat, in order to achieve compost that is sanitary and stable.

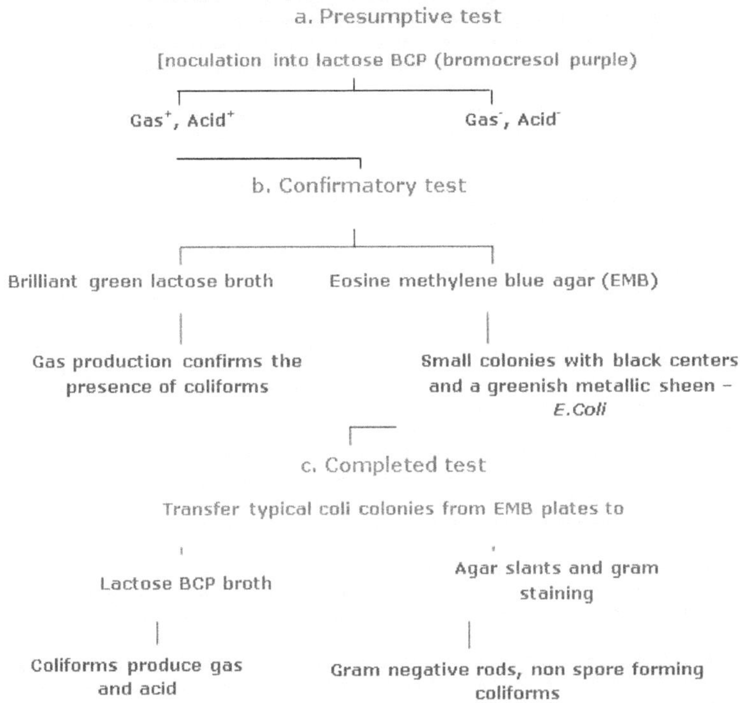

Biowaste (biodegradable waste), any waste that is capable of undergoing anaerobic or aerobic decomposition.

Compost is solid particulate material that is the result of composting, that has been sanitized and stabilized and that confers beneficial effects when added to soil and or used in conjunction with plants.

It is basically a microbiological process accomplished by the combined activity of bacteria, fungi, actinomycetes and protozoa, which are present in the composting material or introduced into the system externally. The purpose of composting wastes ranges from sanitation to volume reduction.

Substrates for composting

Most organic wastes can be successfully composted. Municipal solid wastes, biosolids, animal manures, agricultural, crop and food processing wastes are the excellent candidates for composting. Because they contain carbohydrates, proteins and lipids that the micro organism can easily degrade.

Agricultural wastes

Crop residues like rice straw, wheat straw, sorghum straw, maize, stover, cob, sunflower stalk, sugarcane trash, leaf litter *etc.*, are available in large quantities in the farm.

The agro industry based by products like pressmud, begasse, coir pith, cotton waste, tea waste, mushroom spent waste can be processed for use as manure. According to the type of wastes, the composting process can be adjusted and nutrient rich compost can be prepared.

Classification

Basically composting can be classified based on degree of aeration, temperature and technology. The resulting classes are

(i.) aerobic Vs anaerobic

(ii.) mesophilic Vs thermophilic

(iii.) mechanised Vs non-mechanised systems.

Modern compost systems are aerobic because of the following reasons.

1. aerobic process are not characterised by objectionable odour.
2. elevated temperature may occur in aerobic process, that kills the pathogen and also weed seeds.
3. aerobic composting is more rapid.

Modern composting involves mesophilic followed by thermophilic conditions. Mechanical Vs Windrow compsting is based upon the technology involved in the compost operations. Mechanical composting involves the use of mechanized, enclosed units equipped to provide control of the major environmental factors. Open or windrow composting implies stalking the raw material in elongated piles (windrows) and allowing the composting process to proceed therein.

Composting process

Involves the transformation of the organic component of the solid waste to nutrient rich organic fertilizer. This is effected by microorganism which are influenced by a variety of environmental factors. Mesophilic organisms play a role in initial stages of composting process, latter succession by thermophilic organisms occur. Apart from these, the environmental factors which control the composting process are nutrient content of the substrate, C/N ratio, temperature, pH, aeration, moisture *etc*.

Methods of composting

Major quantities of compost used in agriculture are prepared either by traditional or improved techniques. These cannot be suitably used for processing large quantities of organic waste. Apart from this there is a need to enrich this manure in respect of nitrogen and phosphorus and reduce its bulkiness involving low cost technology. Addition of nutrient source as well as inoculation with microorganism improved the manures.

Microbial inoculants for composting

Naturally occurring organic materials are abundantly colonized with microorganisms to bring about decomposition. But the rate of decomposition will be slower. To hasten the decomposition as well as to increase the rate of decomposition, inoculation is considered essential.

The inocula of selected species of fungi such as Aspergillus, Penicillium, Trichoderma, Paecilomyces, Pleurotus are found to play appreciable role in promoting composting process and reduce the composting period by about four weeks.

Commonly used organisms for decomposing various substrates

Lignin rich waste	-	*Phanirochaete crysosporium* *Tramates versicolor*
Lignocellulosic waste	-	*Pleurotus sojarcaju, Streptomyces*
Cellulosic waste	-	*Cellulomonas, Trichoderma*

Properties of compostable wastes

Certain physical and chemical properties of the organic waste affect the composting process as well as characteristics of the finished product.

Chemical properties

Chemical characteristics may be considered in terms of nutrient quality and quantity. The relative quantity of carbon, N, P, S and other nutrients is important, but the quality of the substance that determines rate of decomposition. Lignin undergoes decomposition much more lower than that of cellulose, the quality of lignin as a C source is much lower than that of cellulose. Maximum rates of decomposition are observed when substrates contain optimum available C, N, P and S ratios. Narrow C/N ratio can lead to loss of N from compost (through ammonia volatalisation). Wider C/N ratio (> 40) in substrates such as oat straw promote immobilization of available N in the compost, resulting in a slowing of decomposition processes due to limited N availability. Addition of exogenous sources of N can overcome such deficiencies and prevent slowing of the composting process.

Physical properties

Most important physical properties of the substrate in relation to composting are particle size and moisture content.

Particle size affects the movement of oxygen into the compost pile and microbial access to substrates. Large particle size promotes oxygen diffusion, but also minimize the specific surface area of the substrate. This means that most of the substrates are not immediately accessible to microbes. Since efficient composting requires access to oxygen and to nutrients in the particles, it is necessary to balance these two requirements.

Optimum moisture content for composting is generally between 40 and 60%. Too much moisture interferes with oxygen accessibility and slows composting. Too little moisture hinders diffusion of the soluble molecules and slows down the composting.

Microbial characteristics of the composting process

Most of the waste carry an indigenous population of microbes from the environment. Three major groups of soil microorganism are normally present when the composting process begins. Microbial population change during the composting process progressing from mesophilic stage into thermophilic stage and then through a gradual cooling period, the final stabilization or curing stage.

Mesophilic microbiota initiate the decomposition of organic substrates and rapidly oxidizes the readily available substrates such as proteins, sugars and starch. Due to microbial metabolism, internal temperature may rise rapidly. The role of mesophile is a preparatory one. They bring the compost into a temperature range suitable for thermophiles by rapidly decomposing the readily available substrates.

In addition to B and F, soil animals such as earthworms and arthropods contribute to the decomposition process by reducing the size of particles greatly increasing the surface area to volume ratio.

Enzymatic activities during composting

Microbes in the compost pile cannot directly metabolize the insoluble particles of organic matter. Rather they produce hydrolytic extracellular enzymes to depolymerize the larger compounds to small fragments that are water soluble. Cellulase activity decreases the amount of cellulose present by about 25% in about 3 weeks. Lipases, proteases and amylase activities rise and fall during the successive stage of composting. Activities of all these enzymes decrease sharply during the thermophilic stage.

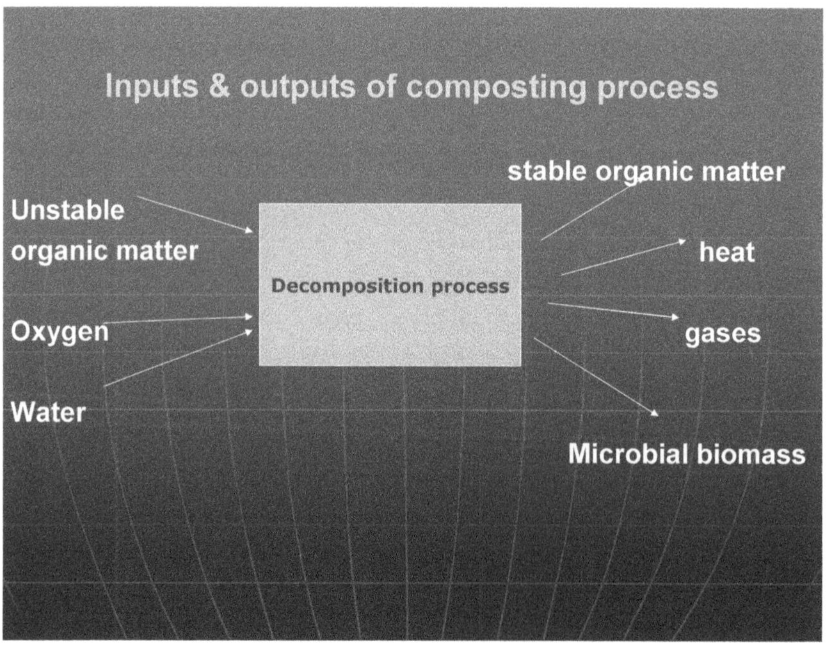

Common bioconversion technology suggested for all the waste materials

- Select a site for composting in an elevated and shady place.
- Select the site in shady place or provide shade.
- Waste can be chopped into 7-10 cm length pieces.
- Spread 100 kg chopped wastes as a bed of 5 x 3 mt.
- Inoculate composite culture.
- Spread another 100 kg chopped waste above this.
- Apply urea according to C : N ratio.
- Form 10 beds similarly.
- Sprinkle sufficient water to maintain moisture of 50 – 60%.
- Cover the surface with tank silt or red soil to give compaction to the bed.
- Once in a week watering should be done with turning.
- Finished compost will be ready within 70 days.
- Due to the biochemical action of lignolytic and cellulolytic microorganism, increase in NPK content as well as considerable reduction in C/N ratio will be observed.

Maturity of the compost can be assessed by

- Colour change from brown to black.
- Volume of the compost heap will be reduced to 1/3.
- Temperature of the heap will be 25-30°C and it should be constant.
- Compost will be very light and fine textured.

10

Soil Microorganisms in Biodegradation of Pesticides and Herbicides

Pesticides are the chemical substances that kill pests and herbicides are the chemicals that kill weeds. In the context of soil, pests are fungi, bacteria insects, worms, and nematodes *etc.* that cause damage to field crops. Thus, in broad sense pesticides are insecticides, fungicides, bactericides, herbicides and nematicides that are used to control or inhibit plant diseases and insect pests. Although wide-scale application of pesticides and herbicides is an essential part of augmenting crop yields; excessive use of these chemicals leads to the microbial imbalance, environmental pollution and health hazards. An ideal pesticide should have the ability to destroy target pest quickly and should be able to degrade non-toxic substances as quickly as possible.

The ultimate "sink" of the pesticides applied in agriculture and public health care is soil. Soil being the storehouse of multitudes of microbes, in quantity and quality, receives the chemicals in various forms and acts as a scavenger of harmful substances. The efficiency and the competence to handle the chemicals vary with the soil and its physical, chemical and biological characteristics.

1. Effects of pesticides: Pesticides reaching the soil in significant quantities have direct effect on soil microbiological aspects, which in turn influence plant growth. Some of the most important effects caused by pesticides are : (1) alterations hi ecological balance of the soil microflora, (2) continued application of large quantities of pesticides may cause ever lasting changes in the soil microflora, (3) adverse effect on soil fertility and crop productivity, (4) inhibition of N2 fixing soil microorganisms such as *Rhizobium, Azotobacter, Azospirillum* etc. and cellulolytic and phosphate solubilizing microorganisms, (5) suppression of nitrifying bacteria, *Nitrosomonas* and *Nitrobacter* by soil fumigants ethylene bromide, Telone, and vapam have also been reported, (6) alterations in nitrogen balance of the soil, (7) interference with ammonification in soil, (8) adverse effect on mycorrhizal symbioses in plants and nodulation in legumes, and (9) alterations in the rhizosphere microflora, both quantitatively and qualitatively.

2. Persistence of pesticides in soil: How long an insecticide, fungicide, or herbicide persists in soil is of great importance in relation to pest management and environmental pollution. Persistence of pesticides in soil for longer period is undesirable because of the reasons: a) accumulation of the chemicals in soil to highly toxic levels, b) may be assimilated by the plants and get accumulated in edible plant products, c) accumulation in the edible portions of the root crops, d) to be get eroded with soil particles and may enter into the water streams, and finally leading to the soil, water and air pollutions. The effective persistence of pesticides in soil varies from a week to several years depending upon structure and properties of the constituents in the pesticide and availability of moisture in soil. For instance, the highly toxic phosphates do not persist for more than three months while chlorinated hydrocarbon insecticides (*e.g.* DOT, aldrin, chlordane *etc*) are known to persist at least for 4-5 years and some times more than 15 years.

From the agricultural point of view, longer persistence of pesticides leading to accumulation of residues in soil may result into the increased absorption of such toxic chemicals by plants to the level at which the consumption of plant products may prove deleterious / hazardous to human beings as well as livestock's. There is a chronic problem of agricultural chemicals, having entered in food chain at highly inadmissible levels in India, Pakistan, Bangladesh and several other developing countries in the world. For example, intensive use of DDT to control insect pests and mercurial fungicides to control diseases in agriculture had been known to persist for longer period and thereby got accumulated in the food chain leading to food contamination and health hazards. Therefore, DDT and mercurial fungicides has been, banned to use in agriculture as well as in public health department.

3. Biodegradation of Pesticides in Soil: Pesticides reaching to the soil are acted upon by several physical, chemical, and biological forces. However, physical and chemical forces are acting upon/degrading the pesticides to some extent, microorganism's plays major role in the degradation of pesticides. Many soil microorganisms have the ability to act upon pesticides and convert them into simpler non-toxic compounds. This process of degradation of pesticides and conversion into non-toxic compounds by microorganisms is known as "biodegradation". Not all pesticides reaching to the soil are biodegradable and such chemicals that show complete resistance to biodegradation are called "recalcitrant".

The chemical reactions leading to biodegradation of pesticides fall into several broad categories which are discussed in brief in the following paragraphs.

a) **Detoxification:** Conversion of the pesticide molecule to a non-toxic compound. Detoxification is not synonymous with degradation. Since a single chance in the side chain of a complex molecule may render the chemical non-toxic.

b) **Degradation:** The breaking down / transformation of a complex substrate into simpler products leading finally to mineralization. Degradation is often considered to be synonymous with mineralization, *e.g.* Thirum (fungicide) is degraded by a strain of *Pseudomonas* and the degradation products are dimethlamine, proteins, sulpholipaids, *etc*.

C. **Conjugation (complex formation or addition reaction):** In which an organism make the substrate more complex or combines the pesticide with cell metabolites. Conjugation or the formation of addition product is accomplished by those organisms catalyzing the reaction of addition of an amino acid, organic acid or methyl crown to the substrate, for *e.g.*, in the microbial metabolism of sodium dimethly dithiocarbamate, the organism combines the fungicide with an amino acid molecule normally present in the cell and thereby inactivate the pesticides/chemical.

d) **Activation:** It is the conversion of non-toxic substrate into a toxic molecule, for *e.g.* Herbicide, 4-butyric acid (2, 4-D B) and the insecticide Phorate are transformed and activated microbiologically in soil to give metabolites that are toxic to weeds and insects.

e) **Changing the spectrum of toxicity:** Some fungicides/pesticides are designed to control one particular group of organisms / pests, but they are metabolized to yield products inhibitory to entirely dissimilar groups of organisms, for *e.g.* the fungicide PCNB fungicide is converted in soil to chlorinated benzoic acids that kill plants.

Biodegradation of pesticides / herbicides is greatly influenced by the soil factors like moisture, temperature, PH and organic matter content, in addition to microbial population and pesticide solubility. Optimum temperature, moisture and organic matter in soil provide congenial environment for the break down or retention of any pesticide added in the soil. Most of the organic pesticides degrade within a short period (3-6 months) under tropical conditions. Metabolic activities of bacteria, fungi and actinomycetes have the significant role in the degradation of pesticides.

4. Criteria for Bioremediation / Biodegradation: For successful biodegradation of pesticide in soil, following aspects must be taken into consideration. i) Organisms must have necessary catabolic activity required for degradation of contaminant at fast rate to bring down the concentration of contaminant, ii) the target contaminant must be bioavailability, iii) soil conditions must be congenial for microbial /plant growth and enzymatic activity and iv) cost of bioremediation must be less than other technologies of removal of contaminants.

According to Gales (1952) principal of microbial infallibility, for every naturally occurring organic compound there is a microbe / enzyme system capable its degradation.

5. Strategies for Bioremediation: For the successful biodegradation / bioremediation of a given contaminant following strategies are needed.

a) **Passive/ intrinsic Bioremediation:** It is the natural bioremediation of contaminant by tile indigenous microorganisms and the rate of degradation is very slow.

b) **Biostimulation:** Practice of addition of nitrogen and phosphorus to stimulate indigenous microorganisms in soil.

c) **Bioventing:** Process/way of Biostimulation by which gases stimulants like oxygen and methane are added or forced into soil to stimulate microbial activity.

d) **Bioaugmentation:** It is the inoculation/introduction of microorganisms in the contaminated site/soil to facilitate biodegradation.

e) **Composting:** Piles of contaminated soils are constructed and treated with aerobic thermophilic microorganisms to degrade contaminants. Periodic physical mixing and moistening of piles are done to promote microbial activity.

f) **Phytoremediation:** Can be achieved directly by planting plants which hyperaccumulate heavy metals or indirectly by plants stimulating microorganisms in the rhizosphere.

g) **Bioremediation:** Process of detoxification of toxic/unwanted chemicals / contaminants in the soil and other environment by using microorganisms.

h) **Mineralization:** Complete conversion of an organic contaminant to its inorganic constituent by a species or group of microorganisms.

11

Biogas Production

BIOGAS (Methane) Biogas is mixture of methane (50-60 per cent), CO_2 (30 – 40 per cent), hydrogen (5-10 per cent), H_2S and nitrogen (traces), produced from anaerobic digestion of animal, plant wastes or any cellulose containing waste material. The digester used for biogas production is called a Biogas plant. A typical biogas plant using cowdung as a raw material consists of: (a) digester and (b) gas holder. The digesters are either of (a) batch type which are filled once, sealed and emptied when the raw materials stop producing gas or (b) continuous type which are fed with a definite quantity of waste at regular intervals so that gas production is continuous and regular. The nature of fermentation in the digester is anaerobic **Methane production** Involves three process *viz.*, hydrolysis, acidification and methonization

Four different groups of organisms are? involved

- **Hydrolytic bacteria** -catabolises carbohydrates,proteins, lipids and other components of biomass into fatty acids,H_2 and CO_2
- **H2 producing acetogenic bacxteria** :Catabolises fatty acids, and the neutral ebd products of I group into acetate,CO_2 and H_2
- **Homoacetogenic bacteria** : Synthesize acetate using H_2,CO_2 and formate
- **Methanogenic bacteria** : Utilizes acetate,CO_2,H_2 to produce methane

The first group of bacteria includes facultative as well as strict anaerobes like Cellulomonas, Bacillus, Eubacteriumetc.

Methanogenic phase

It is a strict anaerobic phase? and during this phase organic Carbon is converted to microbial mass, CO_2 and methane.

These bacteria are sensitive to pH and? optimum is 6.8- 7.2. Drop in pH leads to inhibition of methanogenesis

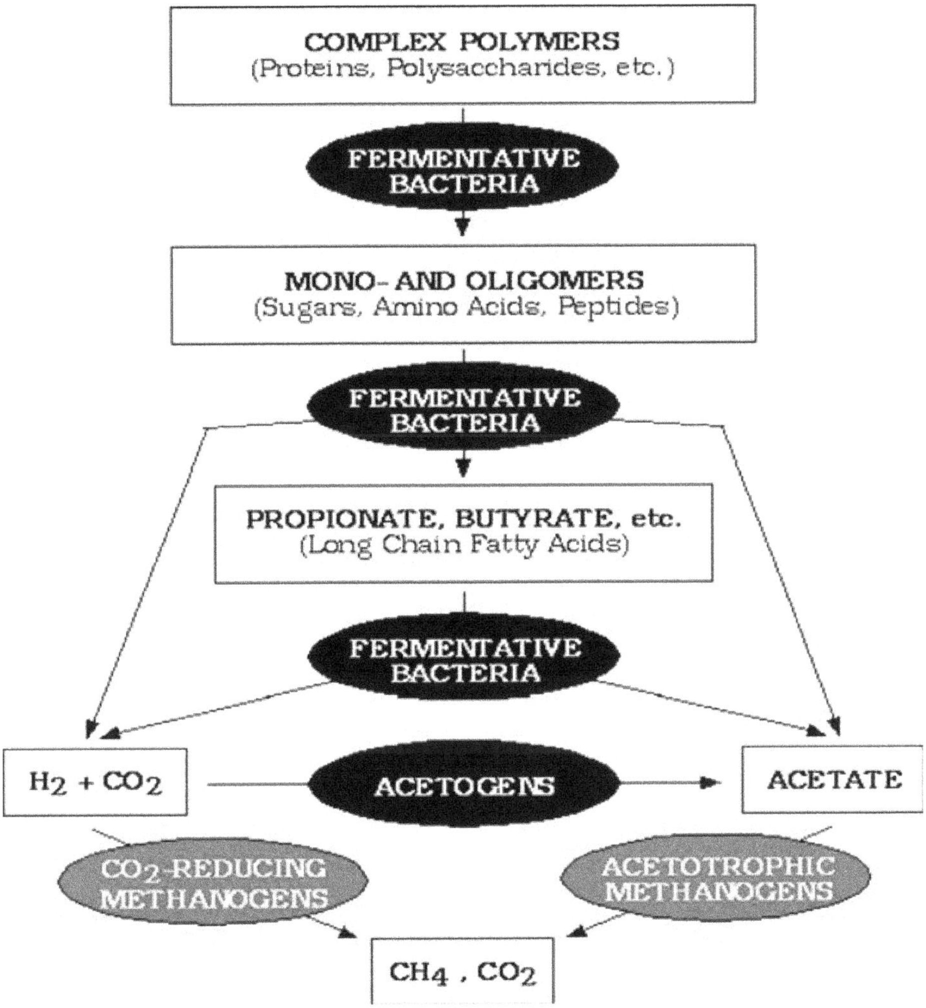
COMPLEX POLYMERS
(Proteins, Polysaccharides, etc.)
FERMENTATIVE BACTERIA
MONO- AND OLIGOMERS
(Sugars, Amino Acids, Peptides)
FERMENTATIVE BACTERIA
PROPIONATE, BUTYRATE, etc.
(Long Chain Fatty Acids)
FERMENTATIVE BACTERIA
H2 + CO2
ACETOGENS
ACETATE
CO2-REDUCING METHANOGENS
ACETOTROPHIC METHANOGENS
CH4 , CO2

12

Biodegradable Plastics

Introduction

Although plastics as we know them today are a relatively recent invention, they have become an important part of modern life. **The age of plastics** Today, 200 billion pounds (100 million tons) of plastics are produced worldwide every year. Plastics are used for packaging, building materials, and virtually every type of consumer product. Past ages of human society have been called the Stone, Bronze, Copper, Iron, and Steel Ages, based on the material that was relied upon the most during that time. Today, the total volume of plastics produced worldwide has surpassed that of steel and continues to increase. Without a doubt, we have entered the Age of Plastics. Some common plastic items include: sunglasses, tooth brushes, super glue, paint brushes, tennis shoes, Frisbees, 2-liter bottles, Honda CRX's, Astroturf, photographs, street signs, pens, automobile paint, video tapes, rubber bands, balloons, bicycle tires, umbrellas, guitar strings, carpeting, shower doors, hearing aids, Scotch Tape, fishing lines, trash bags, and toilet seats. Plastic can be found in everything from clothing to machinary. It is important to understand the nature of plastics, and the consequences of their production and use. Virtually all plastics are made from nonrenewable resources, such as oil, coal or natural gas, which will eventually become exhausted. Plastics waste is increasing, adding to the already burdensome problems of waste management. And the use of plastics continues to grow, raising the important question, how can we balance convenient living with concern for ecology? To understand this concern, it is helpful to understand what plastics are.

Green plastics

Green plastics are the focus of an emerging industry focused on making convenient living consistent with environmental stability. One reason to make a shift toward the use of green plastics is the availability of raw materials. Green plastics can be made using polymers that come from agricultural and marine feedstocks. These are abundant natural resources that are constantly being replenished. This, in turn could revitalize rural economy, both agricultural and marine, by providing additional demand for currently underutilized land or low-valued biomass commodities. Another favorable property of green plastics is their biodegradability, making them a natural material for use in such applications as compostable collection bags, such as for food or yard waste. But bioplastics have to possess adequate physical properties. Their properties have to be managed and controlled with technological means through the development of adequate formulations and plastics processing. The commercial ventures already under way in the United States, Canada, Europe, and Japan indicate that there is confidence technological advances are possible. The key to solving technical problems often simply knows what the problems are. Bioplastics also have to be cost-competitive. Commercially available biopolymers are typically more expensive than synthetic polymers, often significantly so. Currently only starch competes with synthetic polymers in terms of cost. Interest in the development of bioplastics will grow largely to the extent that there is real interest in and concern over the environment. Societal concern over the environment is already being reflected in governmental restrictive legislation on the use of plastics, particularly aimed at plastic packaging. Legislation has begun at the local, state, federal, and international levels, and legislation will undoubtedly increase in the future. New legislation will likely contain restrictions aimed at materials that are neither recyclable nor biodegradable. Labeling legislation may lead to an "ecolabel," based on a product's raw material usage, energy consumption, emissions from manufacture and use, and waste disposal impact. Most of all, what is needed is a paradigm shift.

The widespread use of these new plastics will depend on developing technologies that can be successful in the marketplace. That in turn will partly depend on how strongly society is committed to the concepts of resource conservation, environmental preservation, and sustainable technologies. There are growing signs that people indeed want to live in greater harmony with nature and leave future generations a healthy planet. If so, bioplastics will find a place in the current Age of Plastics.

Plastics

Plastics are a class of material that has one or more polymers as its primary ingredient, that is shaped by flow when it is processed (usually using heat), and that is solid in its final form. Plastics can be made up of many different kinds of polymer, and can be processed in many different ways, but as long as they satisfy these three conditions, they are bona fide plastics. The general "recipe" for any kind of plastic is a combination of three ingredients: a polymer, one or more plasticizers, and one or more additives. These ingredients can then be processed into different shapes, resulting in a wide variety of different materials with different properties.

Polymers

Polymers are long molecules. They are one of the basic components of all plastics.

Synthetic polymers

Synthetic polymers are polymers that are man-made. Most synthetic polymers are manufactured from petroleum. Some examples of synthetic polymers include:

- **Polystyrene** is the polymer found in styrofoam, used for everything from packing materials and insulation to drinking cups.

- **Polyvinyl chloride**, widely known by its abbreviation PVC, is used in a lot of building material (and is well-known as being ubiquitous in piping).

These materials are generally not biodegradable, and because they are made from petroleum, once the basic materials for creating them are used up, we cannot make any more.

Biopolymers

Biopolymers are polymers that occur in nature. Carbohydrates and proteins, for example, are biopolymers. Many biopolymers are already being produced commercially on large scales, although they usually are not used for the production of plastics. Even if only a small percentage of the biopolymers already being produced were used in the production of plastics, it would significantly decrease our dependence on manufactured, non-renewable resources. Some examples of biopolymers include:

- **Cellulose** is the most plentiful carbohydrate in the world; 40 percent of all organic matter is cellulose!
- **Starch** is found in corn (maize), potatoes, wheat, tapioca (cassava), and some other plants. Annual world production of starch is well over 70 billion pounds, with much of it being used for non-food purposes, like making paper, cardboard, textile sizing, and adhesives.
- **Collagen** is the most abundant protein found in mammals. Gelatin is denatured collagen, and is used in sausage casings, capsules for drugs and vitamin preparations, and other miscellaneous industrial applications including photography.
- **Casein**, commercially produced mainly from cow's skimmed milk, is used in adhesives, binders, protective coatings, and other products.
- **Soy protein** and zein (from corn) are abundant plant proteins. They are used for making adhesives and coatings for paper and cardboard.
- **Polyesters** are produced by bacteria, and can be made commercially on large scales through fermentation processes. They are now being used in biomedical applications. A number of other natural materials can be made into polymers that are biodegradable. For example:
- **Lactic acid** is now commercially produced on large scales through the fermentation of sugar feedstocks obtained from sugar beets or sugar cane, or from the conversion of starch from corn, potato peels, or other starch source. It can be polymerized to produce polylactic acid.
- **Triglycerides** can also be polymerized. Triglycerides make up a large part of the storage lipids in animal and plant cells. Over sixteen billion pounds of vegetable oils are produced in the United States each year, mainly from soybean, flax, and rapeseed. Triglycerides are another promising raw material for producing plastics.

These natural raw materials are abundant, renewable, and biodegradable, making them attractive feedstocks for bioplastics.

Plasticizers

A plasticizer is a substance that can be added to material to increase its workability, flexibility, or pliability. Plasticizers are one of basic ingredients of all plastics.

Additives

An additive is a substance that can be added to material to change its properties, usually to make the end product more desirable in some way. Additives are one of basic ingredients of all plastics. The pure polymer resin by itself may not always have the physical properties needed in the final product; it may be strong but too brittle, flexible but too elastic, or flexible and elastic but just plain ugly. Just like the polymer material itself, additives come in different varieties: some can be found in the environment, while others are manufactured. The amounts and types of additives used in manufacturing plastics are another factor that influence how environmentally-friendly they are.

Green Plastics Green Plastics, sometimes also called Bioplastics, are plastics that are biodegradable and are usually made mostly or entirely from renewable resources. Frequently there is also a focus on environmentally friendly processing. Green plastics are the focus of an emerging industry focused on making convenient living consistent with environmental stability. Like all plastics, bioplastics are composed of a polymer, combined with plasticizers and additives, and processed using extrusion or thermosetting. What makes green plastics "green" is one or more of the following properties:

1. they are biodegradable
2. they are made from renewable ingredients
3. they have environmentally friendly processing

Because different compounds can satisfy? some or all of these criteria to different degrees, there are different "degrees of green" in green plastics.

Traditional plastics fail on all three of these points

1. Biodegradability

Biodegradation is a process where something breaks down into simple compounds as a result of the action of microorganisms (like bacteria, fungi, or algae). The term biodegradation is actually a contraction, short for "biotic degradation." Something is biodegradable if it can be broken down by this kind of process. In order to say that something "biodegrades", it therefore has to meet the following requirements:

1. it has to break down (this is simply "degradation")
2. it's molecules have to break down from complex molecules into simpler ones (this is "chemical degradation")

3. The breaking down of its molecules has to be accomplished by microorganisms.
4. In order to prevent misinformation in advertising, standards organizations have made even more strict requirements for something to be labelled as biodegradable. In addition to the above list, something can only be labelled as "biodegradable" if:
5. The biodegradation of the material has to be scientifically measurable. Since most biodegradation produces CO_2 as a by-product, usually this is measured by the amount of CO_2 produced.
6. The biodegradation of the material has to be fast enough to have a significant effect in a reasonable amount of time. For example, the ISO standard requires 60% biodegradation within 180 days for a material to be called biodegradable; the Europenan Norm EN13432 is stricter, requiring 90% biodegradation within 90 days.

Types of Biodegradation

Because biodegradation requires microorganism to do something to a material, usually the material has to be broken up into smaller pieces first. As a result, most biodegradable materials become biodegradable after the action of another kind of degradation.

Hydro-biodegradable

Hydro-biodegradable materials are first broken down by interaction with water (a process called hydrolysis), and then are further broken down by microorganisms.

Photo-biodegradable

Photo-biodegradable materials are first broken down by interaction with sunlight (a process called photolysis), and then are further broken down by microorganisms.

Oxo-degradable

Some companies have been claiming that they have created an additive that can be added to traditional plastics to make them biodegradable. These products become what is called **oxo-degradable**, and sometimes is incorrectly identified as **oxo-biodegradable**. Although this allows the plastic to return to the environment, these products are not biodegradable. Instead, the additive allows the plastic material to break down physically when exposed to water, into pieces small enough to be accidentally ingested by microbes. However, the microbes are not able to actually break this material down further. The end result is therefore a material that combines biomass with polymer residue. The plastic never decomposes as a result of interaction with the organisms. This process is therefore more accurately called "disintergration" rather than "biodegradation".

For bioplastics to become practical, they must have properties that allow them to compete with the current plastics on the market: bioplastics must be able to be strong, resiliant, flexible, elastic, and above all, durable. It is the very durability of traditional plastics that has helped them in the marketplace, and has been a major goal of plastics research throughout the years. However, it is exactly this durability that now has people increasingly worried. Now that we wrap our sandwiches in bags that will still be around when the sandwich, and even the person who ate it, are long gone, many people are wondering: have we gone too far? There is a lot of current research going on concerning methods of decomposition. There is also research on controlling the time-line of biodegradation. One goal of this research is to make a product that is programmed-degradable: in other words, a product that allows you to control when and how it degrades, while insuring that the product remains strong while it is still in use.

2. Renewability

A renewable resource is a natural resource that is created in the environment faster than it is used up by people. Many people think of "renewability" as a fixed trait: some things (like trees, grass, and wind) are renewable, while others (like oil and coal) are not. In fact, whether a resource is renewable depends on both how fast it is replenished and how fast people use it. As a result, some resources are more renewable than others, and some resources may or may not be renewable depending on how they are used.

Rate of Renewal

The rate of rewal (sometimes also called the "sustainable yield") of a resource tells you how quickly it can be replenished by the environment. Solar energy, tides, rainfall, and winds are considered perpetual resources for energy because they renew much faster than they could ever be used. (Can you imagine us "using up" the wind, so that we would have to wait until the earth made more?) Living organisms provide the majority of resources that are generally considered "renewable", because they generally renew themelves within a reasonable amount of time relative to how quickly they are use. Agricultural feedstocks and marine feedstocks are two major categories of living organism feedstocks. Within this category, some organisms renew faster than others: for example, it takes much longer to grow a new tree than it does to grow grass. Most of the resources that are considered "non-renewable" are based on coal, oil, natural gas, and other substances that take so long for the environment to create that almost any use of these resources at all will cause them to be used up before any more is created. Petro-chemical feedstocks are feedstocks derived from petroleum principally for the manufacture of chemicals, synthetic rubber, and a variety of plastics.

Rate of Use

Imagine you live in a small village by a river. A turbine on the river spins, and it can generate enough electricity for the entire village every day. Clearly, their hydroelectric power is a completely renewable resource. However, as the size of the village grows, their energy use grows. If eventually the needs of the village far outstrip the energy that can be provided by the turbine, then the hydroelectric energy from the river is no longer a renewable resource for the village: the rate of use has exceeded the rate of replenishment. The same issue exists for the use of plants. As long as our use of (for example) corn remains moderate compared to the amount of corn produced, corn is a renewable resource. However, if our use of corn increases dramatically without a corresponding increase in corn crop production, then corn will cease to be a renewable resource: we will use it all up, and we will either have to cease production until the corn renews itself or (worse) it will become extinct, so it will not replenish at all. 3. Processing When making plastics, the initial mass of polymer, called resin, is processed into different shapes using a variety of methods, including: extrusion, injection molding, compression molding, transfer molding, and casting. Different processing techniques result in the wide variety of forms that plastic can take: ranging from thin films and elastic sheets, to resiliant panels and hard, solid three-dimensional shapes.

History of Bioplastics

The use of natural polymers is not entirely a new idea. In one form or another, green plastics have been around for a long time.

Early History

Natural resins-like amber, shellac, and gutta percha-have been mentioned throughout history, including during Roman times and the Middle Ages. Native Americans were developing and refining techniques for making ladles and spoons from animal horns long before there was any European contact. In Europe, molded horn jewelry and snuff boxes were popular in the eighteenth century.

The 1800's

Significant commercialization of bioplastics only began in the middle of the nineteenth century... The American inventor, John Wesley Hyatt, Jr., was looking for a substitute for ivory in the manufacture of billiard balls, and in 1869 patented a cellulose derivative for coating non-ivory billiard balls. That attempt, however, was affected by the coating's flammability; balls were occasionally ignited when lit cigars accidentally came into contact with them. Hyatt continued working on the project and soon developed celluloid, the first widely used plastic, now most widely known for its use in photographic and movie film.

The 1900's

The history of plastics changed dramatically in the early 1900s, as petroleum emerged as a source of fuel and of chemicals. The early bioplastics were simply displaced by plastics made from synthetic polymers. World War II brought on a large increase in plastics production, a growth which continues to this day.

The 1920's

In the 1920s Henry Ford experimented with using soybeans in the manufacture of automobiles. Ford was partly motivated by a desire to find non-food applications for agricultural surpluses, which existed then as they do now. Soy plastics were used for an increasing number of automobile parts, like steering wheels, interior trim, and dashboard panels. Finally Ford gave the go-ahead to produce a complete prototype "plastic car." Ford, a master at generating publicity, exhibited the prototype with great fanfare in 1941, but by the end of the year was no longer publicizing the "plastic car," probably for a variety of reasons. World War II played a role: armament work took precedent over almost everything else, and steel shortages limited all non-defense production. Today plastic automobile parts are common, but the use of plastics made from renewable raw materials got side-tracked.

The 1960's

One well established bioplastic that has survived the growth of the synthetic plastics industry is cellophane, a sheet material derived from cellulose. Although production peaked in the 1960s it is still used in packaging for candy, cigarettes, and other articles.

The 2000's and Beyond

Demand for materials like plastics is continually growing and will not be abated. Today, the plastics industry is an important component of our economy: The U.S. plastics industry includes over 20,000 facilities that produce or distribute materials or products, employ over 1.5 million workers, and ship over $300 billion in products each year. The magnitude of the plastics industry, however, is itself a cause for concern. The pressures of increasing waste and diminishing resources have lead many to try to re-discover natural polymers and put them to use as materials for manufactor and industry. As a result, there is increasing interest in the promise of a new generation of green plastics.

13

Bioremediation

Introduction

In the twentieth century, the ever increase in the global human population and industrialization led to the exploitation of natural resources. The increased usage of heavy metals and its disposal in the ecosystems around the world and also in India is becoming an important ecological issue to be take into our consideration. This has indiscriminate release of heavy metals into the ecosystem has already posing air, water and soil pollution causing various uncompromising, deleterious and fatal effects on humans and the stability of the ecosystems. Soil and water are the most important natural resources for all living beings including human beings for survival, which are recently becoming highly polluted. As a result of this, several living forms which are better suited to the polluted environment have out bursted causing ecological imbalance. Unlike soil, water body has its own ability to maintain its natural state through process called self purification. However when the discharge of pollutants is heavy, the process of self purification of water bodies is adversely affected and the water remains polluted. Unlike organic contaminants, heavy metals are not biologically degradable, and therefore can persist in the environment for a long duration. The term heavy metal' can be explained as 1) Relatively abundant in the earth's crust; 2) Reasonable extraction and usage; 3) Having direct contact with people; and 4) Toxic to humans. Heavy metals are the metals which have a specific gravity of more than 4 or more than 5 (Anonymous, 1964; Nieboer and Richardson, 1980). Some important heavy metals are Zinc (Zn), Chromium (Cr), Cadmium (Cd), Arsenic (As) *etc.*, Bioremediation can be defined as any process that uses microorganisms, fungi, green plantsor their enzymes to return the natural environment altered by contaminants to its original condition. Bioremediation may

be employed to attack specific soil contaminants, such as degradation of chlorinated hydrocarbons by bacteria. An example of a more general approach is the cleanup of oil spills by the addition of nitrate and/or sulfate fertilisers to facilitate the decomposition of crude oil by indigenous or exogenous bacteria. Mycoremediation is a form of bioremediation that uses fungi to reduce the level of contamination in a given environment. Fungi can release specific enzymes and acids that break down the major components of plant fiber. The accumulation of waste is directly proportional to population and as our population grows scientists will be under more pressure to find ways to eliminate contaminants from our environment. Bioremediation microbes will continue to be used in an attempt to return our polluted environments to their original state. Corning, Barnstead, and BD supply products used by those researching bioremediation.

Bioremediation technologies can be generally classified as in situ or ex situ. In situ bioremediation involves treating the contaminated material at the site while ex situinvolves the removal of the contaminated material to be treated elsewhere. Some examples of bioremediation technologies are bioventing, landfarming, bioreactor, Composting, bioaugumentation, rhizofiltration, and biostimulation

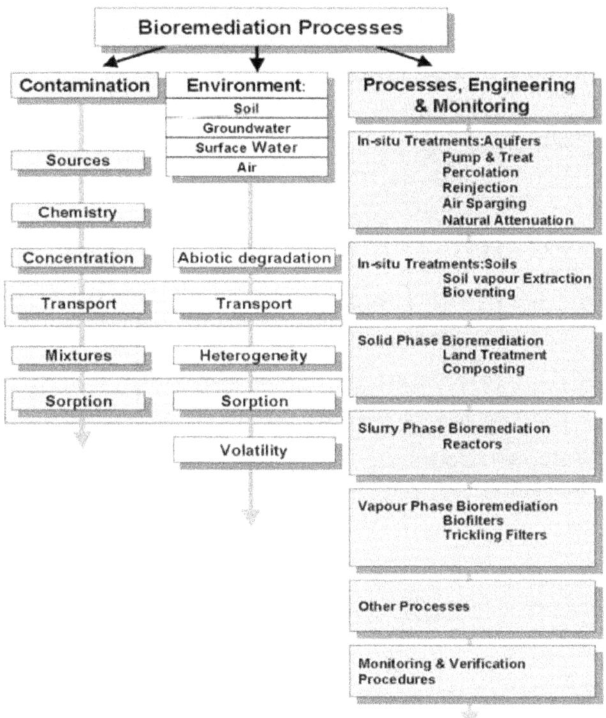

BIOREMEDIATION STRATEGIES Different techniques are employed depending on the degree of saturation and aeration of an area. In situ techniques are defined as those that are applied to soil and groundwater at the site with minimal disturbance. Ex situ techniques are those that are applied to soil and groundwater at the site which has been removed from the site via excavation (soil) or pumping (water). Bioaugmentation techniques involve the addition of microorganisms with the ability to degrade pollutants.

In situ **bioremediation** These techniques are generally the most desirable options due to lower cost and less disturbance since they provide the treatment in place avoiding excavation and transport of contaminants. In situtreatment is limited by the depth of the soil that can be effectively treated. In many soils effective oxygen diffusion for desirable rates of bioremediation extend to a range of only a few centimeters to about 30 cm into the soil, although depths of 60 cm and greater have been effectively treated in s ome cases. The most important land treatments are:

- **Incineration**
- **Excavation**

Bioventing Is the most common in situ treatment and involves supplying air and nutrients through wells to contaminated soil to stimulate the indigenous bacteria. Bioventing employs low airflow rates and provides only the amount of oxygen necessary for the biodegradation while minimizing volatilization and release of contaminants to the atmosphere. It works for simple hydrocarbons and can be used where the contamination is deep under the surface.

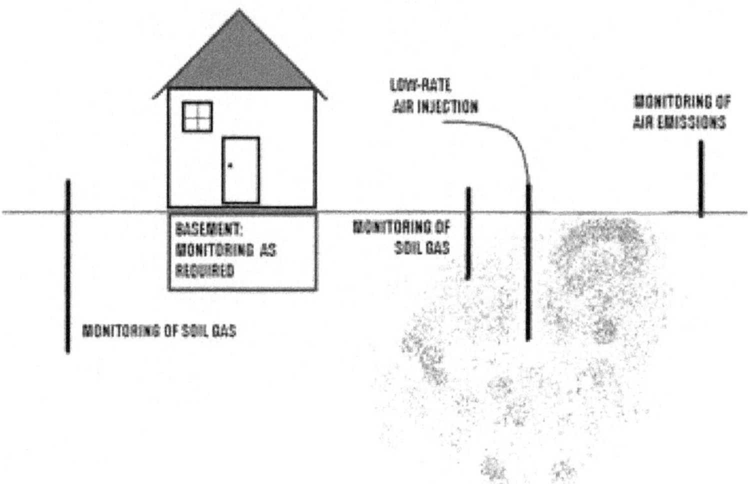

In situ biodegradation Involves supplying oxygen and nutrients by circulating aqueous solutions through contaminated soils to stimulate naturally occurring bacteria to degrade organic contaminants. It can be used for soil and groundwater. Generally, this technique includes conditions such as the infiltration of water-containing nutrients and oxygen or other electron acceptors for groundwater treatment.

Biosparging: Biosparging involves the injection of air under pressure below the water table to increase groundwater oxygen concentrations and enhance the rate of biological degradation of contaminants by naturally occurring bacteria. Biosparging increases the mixing in the saturated zone and there- by increases the contact between soil and groundwater. The ease and low cost of installing small-diam- eter air injection points allows considerable flexibility in the design and construction of the system.

Bioaugmentation: Bioremediation frequently involves the addition of microorganisms indigenous or exogenous to the contaminated sites. Two factors limit the use of added microbial cultures in a land treatment unit: 1) nonindigenous cultures rarely compete well enough with an indigenous population to develop and sustain useful population levels and 2) most soils with long-term exposure to biodegradable waste have indigenous microorganisms that are effective degrades if the land treatment unit is well managed.

Ex-situ bioremediation

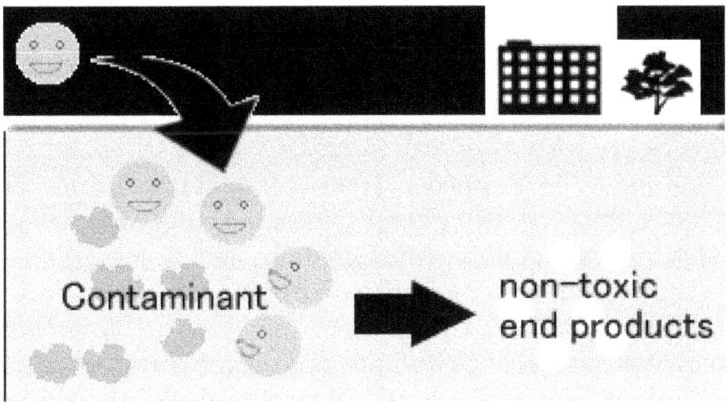

These techniques involve the excavation or removal of contaminated soil from ground.

Landfarming Is a simple technique in which contaminated soil is excavated and spread over a prepared bed and periodically tilled until pollutants are degraded. The goal is to stimulate indigenous biodegradative microorganisms and facilitate their aerobic degradation of contaminants. In general, the practice is limited to the treatment of superficial 10–35 cm of soil. Since landfarming has the potential to reduce monitoring and maintenance costs, as well as clean-up liabilities, it has received much attention as a disposal alternative.

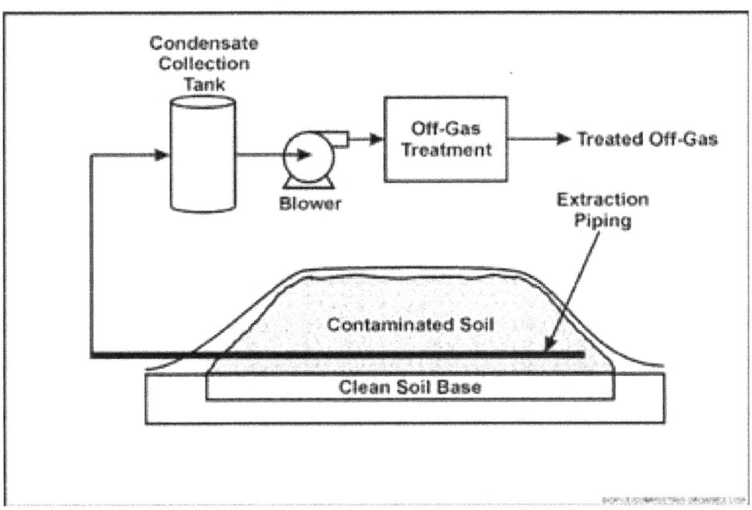

Composting Is a technique that involves combining contaminated soil with nonhazardous organic amendants such as manure or agricultural wastes. The presence of these organic materials supports the development of a rich microbial population and elevated temperature characteristic of composting.

Biopile Are a hybrid of landfarming and composting. Essentially, engineered cells are constructed as aerated composted piles. Typically used for treatment of surface contamination with petro- leum hydrocarbons they are a refined version of landfarming that tend to control physical losses of the contaminants by leaching and volatilization. Biopiles provide a favorable environment for indigenous aerobic and anaerobic microorganisms.

Bioreactors Slurry reactors or aqueous reactors are used for ex situ treatment of contaminated soil and water pumped up from a contaminated plume. Bioremediation in reactors involves the processing of contaminated solid material (soil, sediment, sludge) or water through an engineered containment system. A slurry bioreactor may be defined as a containment vessel and apparatus used to create a three-phase (solid, liquid, and gas) mixing condition to increase the bioremediation rate of soil- bound and water-soluble pollutants as a water slurry of the contaminated soil and biomass (usually indigenous microorganisms) capable of degrading target contaminants. In general, the rate and extent of biodegradation are greater in a bioreactor system than in situ or in solid-phase systems because the contained environment is more manageable and hence more controllable and predictable. Despite the advantages of reactor systems, there are some disadvantages. The contaminated soil requires pre treatment (*e.g.*, excavation) or alternatively the contaminant can be stripped from the soil via soil washing or physical extraction (*e.g.*, vacuum extraction) before being placed in a bioreactor.

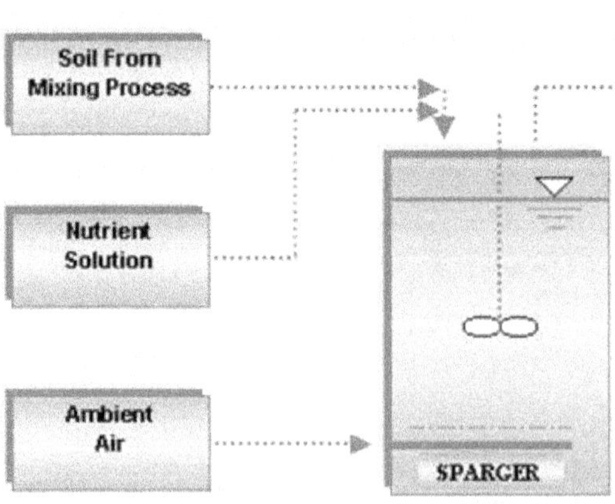

Stirred Batch Reactor

Advantages of bioremediation

- Bioremediation is a natural process and is therefore perceived by the public as an acceptable waste treatment process for contaminated material such as soil. Microbes able to degrade the contaminant increase in numbers when the contaminant is present; when the contaminant is degraded, the biodegradative population declines. The residues for the treatment are usually harmless products and include carbon dioxide, water, and cell biomass.

- Theoretically, bioremediation is useful for the complete destruction of a wide variety of contaminants. Many compounds that are legally considered to be hazardous can be transformed to harmless products. This eliminates the chance of future liability associated with treatment and disposal of contaminated material.
- Instead of transferring contaminants from one environmental medium to another, for example, from land to water or air, the complete destruction of target pollutants is possible.
- Bioremediation can often be carried out on site, often without causing a major disruption of nor- mal activities. This also eliminates the need to transport quantities of waste off site and the poten- tial threats to human health and the environment that can arise during transportation.
- Bioremediation can prove less expensive than other technologies that are used for clean-up of hazardous waste.

Disadvantages of bioremediation

- Bioremediation is limited to those compounds that are biodegradable. compounds are susceptible to rapid and complete degradation.
- There are some concerns that the products of biodegradation may be more persistent or toxic than the parent compound.
- Biological processes are often highly specific. Important site factors required for success include the presence of metabolically capable microbial populations, suitable environmental growth conditions, and appropriate levels of nutrients and contaminants.
- It is difficult to extrapolate from bench and pilot-scale studies to full-scale field operations.
- Research is needed to develop and engineer bioremediation technologies that are appropriate for sites with complex mixtures of contaminants that are not evenly dispersed in the environment. Contaminants may be present as solids, liquids, and gases.
- Bioremediation often takes longer than other treatment options, such as excavation and removal of soil or incineration.
- Regulatory uncertainty remains regarding acceptable performance criteria for bioremediation. There is no accepted definition of clean, evaluating performance of bioremediation is difficult, and there are no acceptable endpoints for bioremediation treatments.

Unit III: Food Microbiology

14

Microorganisms Associated with Food & Dairy Products

I. Microorganisms and food

Microorganisms associated in a variety of ways with all food we eat. Food consumed by man and animals are ideal ecosystems in which bacteria and fungi can multiply. The more presence of microorganisms in food in small numbers need not be harmful, but their unrestricted growth may render the food unfit for consumption and can result in spoilage or deterioration. Some organisms can grow and produce secondary metabolites, that affect the food quality, which may be either desirable or undesirable *e.g.* Lactic acid production and also acetification of wine. Some organisms, not only causes spoilage but also produce metabolites which are toxic to man and animals *e.g.* Aflatoxin by aspergilli in feed. Some microorganisms are used in preparation and preservation of food products *e.g.* Yogurt and Sauerkraut.

II. Microorganisms associated with food products

All foods carry contaminating organisms from natural sources. Contamination begins from the start of handling by humans and it continues till the product is consumed.

1. Cereals

Cereals and cereal products which include grain, flours, bread and other bakery products. They contain organisms from soil, insects and other sources. Bacteria found on freshly harvested grains include species of Pseudomonas, Micrococcus, Lactobacillus, Bacillus *etc.*, and the total number ranges from a few to several millions. Most of these are removed during milling.

In wheat flour bacteria occurring are species of Bacillus, Coliforms, Micrococcus, Sarcina, Serratia *etc.*, fungi include Aspergillus and Penicillium. The surface of the freshly baked bread is free of viable M.o but may get contaminated during cooling and wrapping.

Spoilage

In cereals and cereals products, when moisture increases > 20%, molds and bacteria grow and cause spoilage. Moisture, physical damage and temperature are the major factors involved in mold spoilage. Moldy growth may lead to mycotoxin production also. Moist flour will allow mold growth, wet mash will undergo acid fermentation by coliforms and lactic bacteria. As soon as acidity increases, yeast growth will be favoured.

Spoilage organisms of baked products

Rhizopus nigricans, Penicillium expansum, Aspergillus niger, Bacillus subtilis, B. licheniformis *etc.*

2. Sugar and Sugar products

Syrups, jam and jellies are the products mainly contaminated with bacteria – Pseudomonas, Flavobacterium, Leuconostoc, Bacillus and the yeasts Saccharomyces, Candida and Pichia. Honey contain yeast, rarely contain bacteria. Because of high sugar content (70-80%) and low pH (3.2-4.2), mostly spoiled by osmophilic yeasts (Zygosaccharomyces / Torula) and the spoilage result in the production of CO_2, alcohol and non volatile acids, which give an off flavour to honey. Candies are not easily contaminated, the soft fillings in chocolate covered candies may support microbial growth.

3. Fruits and vegetables

Fruits and vegetables are subject to contamination both in the field as well as during harvesting and storage. Susceptible to infection by bacteria, fungi and viruses. Bacterial genera present on fruits and vegetables include – Pseudomonas, Bacillus, Enterobacter, Sarcina, Staphylococcus, Lactobacillus, Leuconostoc. In fruits pH is acidic bacterial growth is restricted and vegetables are susceptible to bacterial attack. Growth of these organisms may take place between harvesting and processing.

Common type of microbial spoilage

Bacterial soft rot	-	*Erwinia carotovora* – degrades pectin and gives softy appearance and bad odour
Grey mold rot	-	*Botrytis* species
Rhizopus soft rot	-	*Rhizopus* species with cotton growth leads to softening
Sliminess or society	-	Saprophytic bacteria

Bacterial soft rot is common in non acid vegetables and fruits. *Rhizopus* rot occur in acidic vegetables and fruits. Bacterial spoilage occurs in low acid fruit juices.

4. Meat and meat products

Meat and meat products are the excellent growth media for variety of bacteria. Though outer part contains more Microorganism, inner side, it is less, the contamination comes from external sources.

Molds	-	*Cladosporium, Sporotrichum, Geotrichum, Penicillium, Mucor*
Bacteria	-	*Pseudomonas, Micrococcus, Streptococcus, Lactobacillus, Salmonella, Escherichia, Clostridium*

Raw meat is subjected to spoilage by its own enzyme as well as by microbial action.

Aerobic spoilage	-	Surface slime by *Pseudomonas, Streptococcus, Leuconostoc*
Yeast and mold	-	causes stickeyness, rancid and spotty
Anaerobic spoilage	-	Facultative and anaerobic bacteria – sowing and putrefaction

5. Fish

The microflora of the living fish depends the microflora of the water. Fish microflora includes *Pseudomonas, Alcaligenes, Vibrio, Serratia, Micrococcus*. Bacteria causing spoilage of fish are a part of natural flora and the predominent spoilage bacteria are *Pseudomonas, Achromobacter, Flavobacterium, Micrococcus* and *Bacillus*.

6. Eggs and Poultry

Freshly laid eggs are sterile, can carry microorganism from faecal matter and from the environment. Microorganism enters through cracks in the shells. Streptococcus, Staphylococcus, Sarcina, Micrococcus are recovered from egg shell.

Spoilage

Green rot	-	*P. fluorescens*
Black rot	-	*Proteus, Pseudomonas*
Colorless rot	-	*Coliforms, Alcaligenes*
Molds spoilage	-	*Penicillium, Cladosporium, Sporotrichum*

Poultry

Pseudomonas are major contaminants on the skin of freshly dressed poultry. Under good sanitation condition, bacterial count is 100 – 1000/sq.cm skin surface. Major bacteria on poultry and poultry products are *Enterobacter, Escherichia, Staphylococcus, Proteus* and *Salmonella*.

III. Micro organisms associated with Dairy Products

Milk, as a complete food, it is a good medium for the growth of micro organism. It contains very few bacteria at secretion. During milking – exterior of the udder, dairy utensils, milking machines, hands of milking man and from soil and dust. Bacteria yeast and mold get into milk and constitute normal flora of the milk. Presence of non-pathogenic microorganism is not serious and excess multiplication causes spoilage (Souring, putrefaction and develop undesirable odour). Pathogenic microorganism include *Streptococcus cremoris, Pseudomonas* sp., *Mycobacterium, Serratia marcescens, Coxiella burnetit,* enteric bacteria *etc.*

Spoilage of milk and milk products

i. Curdling:

- Lactic acid bacteria, coliforms, *Micrococcus, Bacillus* .

ii. Acid production –

- At high temperature (37-50 oC), Streptococcus *thermophilus, S. faecalis, Lactobacillus bulgaricus*- can grow & produce lactic acid.
- Pasteurization kills acid producing bacteria & may permit survival of heat Rt lactics- may cause lactic fermentation.
- Coliforms also produce- mixture of acids (lactic, acetic, formic)

iii. Gas production –

- H_2 & Co_2 – by *Clostridia, Bacilli,* yeast & heterolactis- foam production
- Curd- ripped off due to excess gas called **stromy fermentation** of milk.

IV. Proteolysis

- Proteins are hydrolysed with the production of bitter taste. (proteolytic M.O)
- *Micrococcus, Alcaligenes, Pseudomonas, Proteus, Clostridium.*
- Acid proteolysis, sweet curdling- slow proteolysis, proteolysis with alkalinity.

V. Roppiness-

- **Bacterial roppiness-** growth & gum production by *Alcaligenes viscolactis, Micrococcus, Streptococcus.*

VI. Colour change

- Variety of bacteria can cause colour change.
- *Pseudomonas*- blackish grey
- *Flavobacterium*-yellow colour
- *Serratia* – red pigments

VII. Rancidity

- Lipolytic organisms causes rancidity by the oxidation of unsaturated fatty acids –by lipases.
- *Alcaligenes*, Bacillus, Proteus, Micrococcus.

Fermented dairy products

Product	Organism
Cultured butter milk	*Streptococcui, & leuconostoc*
Cultured sour cream	*Streptococcui, & leuconostoc*

Product	Organism
Bulgarian milk	*L.bulgaricus*
Acidophilus milk	*L.acidophilus*
yogurt	*S.thermophilus, L.bulgaricus*
Kefir	*S. lactis, L.bulgaricus*
Cottage cheese	*S. lactis*
Cheddar, swiss chesse	*S.faecalis*
Cambert- soft cheese	*Penicillium caneberti, Brevibacterium*
Roquefort- semi soft	*Penicillium roqueforti*

IV. Food infections and Food intoxications

A bacterial food intoxication refers to illness caused by the consumption of a bacterial toxin formed in the food, while a bacterial food infection refers to illness caused by the entrance of food and reaction of the body to their presence or to their metabolites.

Food intoxications

Two major food intoxications are caused by bacteria.

i. Botulism

ii. Staphylococcal intoxication.

Food infections and intoxications

- **Food intoxication** – illness caused by the consumption of micrbial toxins in the food
- **Food infection** - – illness caused by the entrance M.O into the body through contaminated food
- **Major food intoxication**
- **Botulism** –caused by the presence of toxin produced by Clostridium botulinum in food
- Neurotoxin & sources of botulism are canned meat, fish, sweet corn,beets and other low –medium acid foods.
- Symptoms-digeastive disturbences, nausea,vomitting, diarrhoea with dizziness & headache.
- **Staphylococcal intoxication**
- Enterotoxin produced by Staphylococcus aureus
- Illness starts after 2-6 hrs of ingestion symptoms nausea, vomitting abdominal pain & diarrhoea

Food infections

- *Salmonellosis*
- *Clostridium perfringens* illness
- *Bacillus cereus gasteroenterities*
- *E. coli* infection
- Shigellosis

Unit IV: Microbial Biotechnology

15

Biofertilizers

Introduction to Biofertilizers

Biofertilizers are microbial inoculants or carrier based preparations containing living or latent cells of efficient strains of nitrogen fixing, phosphate is solublizing and cellulose decomposing microorganisms intended for seed or soil application and designed to improve soil fertility and plant growth by increasing the number and biological activity of beneficial microorganisms in the soil.

The objects behind the application of Biofertilizers /microbial inoculants to seed, soil or compost pit is to increase the number and biological / metabolic activity of useful microorganisms that accelerate certain microbial processes to augment the extent of availability of nutrients in the available forms which can be easily assimilated by plants. The need for the use of Biofertilizers has arisen primarily due to two reasons *i.e.* though chemical fertilizers increase soil fertility, crop productivity and production, but increased / intensive use of chemical fertilizers has caused serious concern of soil texture, soil fertility and other environmental problems, use of Biofertilizers is both economical as well as environment friendly. Therefore, an integrated approach of applying both chemical fertilizers and Biofertilizers is the best way of integrated nutrient supply in agriculture.

Organic fertilizers (manure, compost, vermicompost) are also considered as Biofertilizers, which are rendered in available forms due to the interactions of microorganisms or their association with plants. Biofertilizers, thus include i) Symbiotic nitrogen fixers *Rhizobium sp.* ii) Non-symbiotic, free living nitrogen fixers *Azotobacter, Azospirillum* etc. iii) BGA-inoculants *Azolla-Anabaena,* iv) Phosphate solubilizing microorganisms (PSM) *Bacillus Pseudomonas, Penicillium Aspergillus* etc. v) Mycorrhiza vi) Cellulolytic microorganisms and vii) Organic fertilizers.

Nobbe and Hiltner (1895, USA) produced the first *Rhizobium* biofertilizer under the brand name "Nitragin" for 17 different legumes.

Role of Biofertilizers in soil fertility and Agriculture

Biofertilizers are known to play a number of vital roles in soil fertility, crop productivity and production in agriculture as they are eco friendly and can not at any cost replace chemical fertilizers that are indispensable for getting maximum crop yields. Some of the important functions or roles of Biofertilizers in agriculture are:

1. They supplement chemical fertilizers for meeting the integrated nutrient demand of the crops.
2. They can add 20-200 kg N/ha year (*e.g.* Rhizobium sp 50-100 kg N/ha year ; Azospirillum , Azotobacter : 20-40 kg N/ha /yr; Azolla : 40-80 kg N/ha; BGA :20-30 kg N/ha) under optimum soil conditions and thereby increases 15-25 percent of total crop yield.
3. They can at best minimize the use of chemical fertilizers not exceeding 40-50 kg N/ha under ideal agronomic and pest-free conditions.
4. Application of Biofertilizers results in increased mineral and water uptake, root development, vegetative growth and nitrogen fixation.
5. Some Biofertilizers (eg, Rhizobium BGA, Azotobacter sp) stimulate production of growth promoting substance like vitamin-B complex, Indole acetic acid (IAA) and Gibberellic acids *etc.*
6. Phosphate mobilizing or phosphorus solubilizing Biofertilizers / microorganisms (bacteria, fungi, mycorrhiza *etc.*) converts insoluble soil phosphate into soluble forms by secreting several organic acids and under optimum conditions they can solubilize / mobilize about 30-50 kg P2O5/ha due to which crop yield may increase by 10 to 20%.
7. Mycorrhiza or VA-mycorrhiza (VAM fungi) when used as Biofertilizers enhance uptake of P, Zn, S and water, leading to uniform crop growth and increased yield and also enhance resistance to root diseases and improve hardiness of transplant stock.
8. They liberate growth promoting substances and vitamins and help to maintain soil fertility.
9. They act as antagonists and suppress the incidence of soil borne plant pathogens and thus, help in the bio-control of diseases.
10. Nitrogen fixing, phosphate mobilizing and cellulolytic microorganisms in biofertilizer enhance the availability of plant nutrients in the soil and thus, sustain the agricultural production and farming system.
11. They are cheaper, pollution free and renewable energy sources
12. They improve physical properties of soil, soil tilth and soil health in general.
13. They improve soil fertility and soil productivity.
14. Blue green algae like Nostoc, Anabaena, and Scytonema are often employed in the reclamation of alkaline soils.
15. Bio-inoculants containing cellulolytic and lignolytic microorganisms enhance the degradation/ decomposition of organic matter in soil, as well as enhance the rate of decomposition in compost pit.
16. BGA plays a vital role in the nitrogen economy of rice fields in tropical regions.
17. Azotobacter inoculants when applied to many non-leguminous crop plants, promote seed germination and initial vigor of plants by producing growth promoting substances.

18. Azolla-Anabaena grows profusely as a floating plant in the flooded rice fields and can fix 100-150 kg N/ha /year in approximately 40-60 tones of biomass produced,
19. Plays important role in the recycling of plant nutrients.

Biofertilizers

Biofertilizers are the preparations containing live or latent cells of efficient strains of nitrogen fixing, phosphate solubilizing or cellulolyotic micro-organisms used for application to seed or composting areas with the objective of increasing the numbers of such micro-organisms and accelerating those microbial processes which augment the availability of nutrients that can be easily assimilated by plants. Biofertilizers harness atmospheric nitrogen with the help of specialized micro-organisms which may be free living in soil or symbiotic with plants. 'Microbial inoculants' are carrier based preparations containing beneficial micro-organisms in a viable state, intended for seed or soil application, designed to improve soil fertility and help plant growth by increasing the number of desired micro-organisms in plant rhizosphere.

Microbial Inoculants: - In soil the activities of Nitrogen fixation, mobilisation of plant nutrients and degradation of ligno-cellulotic wastes are being carried out by a large number of micro-organisms. Artificially multiplied cultures of selected micro organisms augment the natural recycling of organic resources. There are different types of microbial inoculants.

Rhizobium

Legumes provide a protein – rich source of food, which is an essential part of the diet in many parts of the tropics. They play an important role in supplying those essential amino acids, particularly lysine that are not present in sufficient quantities in cereal crops. These leguminous plants in association with the soil bacterium, Rhizobium form root nodules and fix atmospheric N_2.

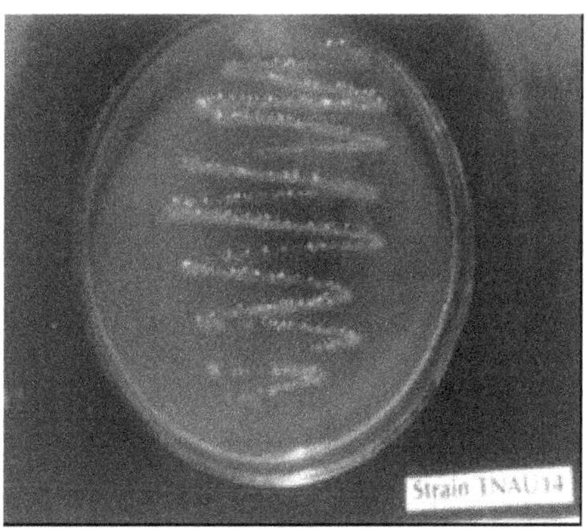

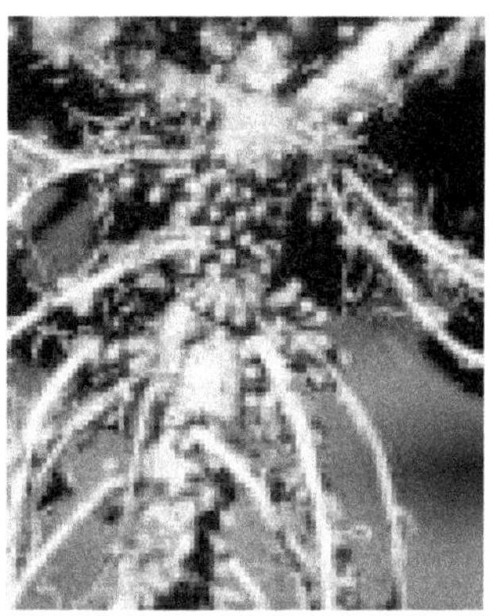

The rhizobia that form associations with the best know legumes have been characterized by nucleic acid sequence homologies as well as by the more traditional traits such as growth rates (fast and slow), acid production from carbohydrates, serology and legume associations. The success of these legume crops depends upon the efficient functioning of legume plant – Rhizobium symbiosis. The efficient functioning of nitrogen fixation under field conditions will contribute to improving the yield of legume crops. Yield can be influenced by several factors including the genotype of the plant, native rhizobial population, quality of rhizobial inoculant, and the environmental and ecological cond*itions in which legume crops are grown.*

Cross inoculation groups of rhizobium (CIG)

It (CIG) refers the groups of leguminous plants that will develop effective nodules when inoculated with the rhizobia obtained from the nodules from any member of that legume group.

I. Rhizobium

1.	Rhizobium leguminosarum	CIG	Host it can nodulate
	bv. viceae	Pea	Peas, lenfils, vicia
	bv. phaseoli	Bean	Phaseolus spp
	bv. trifoli	Clover	Trifolium spp
2.	*R. meliloti*	Alfalfa	Alfalfa, clover, fenugreek
3.	*R. loti*	Lotus	Trifoli, lupine,
4.	*R. fredii*	Soybean	Soybean
5.	*R. spp*	Cowpea group	Vigna, Arachis, Cajanus, Dolichus, Sesbania, Acacia, Prosopis, green gram and blackgram
6.	*R. sp*	Chickpea group	Chickpea

II. Bradyrhizobium

B. japonicum	Soybean	Soybean
B. spp	Cowpea group	Cowpea group plants

III. Azorhizobium

Stem nodulating – one. Nodulates Sesbania rostrata.

IV. Photorhizobia

Nodulants aeschynomene sp.

V. Sinorhizobium

Fast growing soybean nodulator

To ensure the presence of adequate numbers of rhizobia in the root rhizosphere of legume crops, these are inoculated with superior rhizobial strains. The success or failure of rhizobial inoculants is largely governed by the combined interaction of various biotic and abiotic factors. Hence a large-scale screening should be carried out to identify ideal inoculant strain for different legume crops. The criterion for selection of efficient strain may vary for particular soil types like acidic, sodic-saline, nitrate rich or heavy metal contamination.

Azospirillum

A number of free living bacteria belonging to the genera Azotobacter, Azospirillum, Beijerinckia, Klebsiella, Bacillus, Pseudomonas, *etc*. have been reported to fix atmospheric nitrogen. Due to some reason or other work connected with most of the above mentioned organisms remained within the walls of the laboratories and never reached the field. In India Azotobacter has been tried as a biofertilizer for crops like rice and cotton. The problem faced with Azotobacter is that substantial amounts of energy is required by Azotobacter for nitrogen fixation. The possible source of energy is the soil organic matter. The energy generated during the decomposition of organic matter is used for nitrogen fixation. In soils where the level of organic matter is very low Azotobacter runs short of energy source and ceases to fix nitrogen.

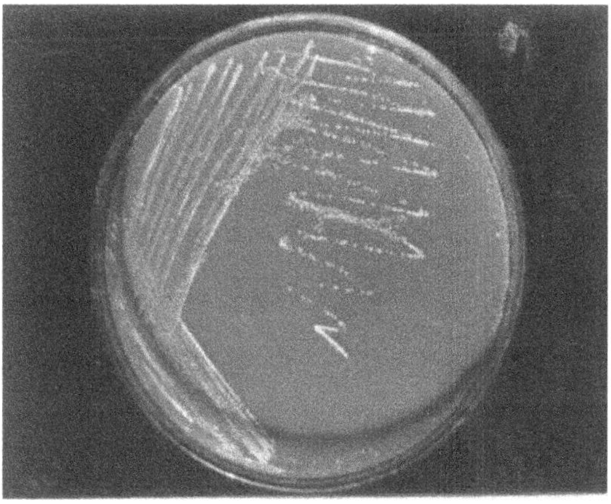

The recognition in the mid seventies that strains of Azospirillum are abundantly associated with the roots of cereals and millets and help fixing substantial amounts of nitrogen led to the opening of the new line of work, the associative symbiosis. Azospirillum is considered to be an associative symbiotic bacterium because it lives in close association with the root system of plants. Almost all crop plants harbour this bacterium in their roots. These bacteria colonising the roots not only remain on the root surface but also a sizable population penetrates into the root tissues and live in harmony with the plant. However they do not produce any visible nodules or outgrowth on the roots. In the absence of any plant, Azospirillum lives in the soil just like any other microorganism saprophytically. However, when a suitable crop is raised they are attracted towards the root system. Making use of the nutrients exuded by the plants through their root system (root exudates) Azospirillum colonies in the root region in large numbers and fixes substantial amounts of nitrogen.

All azospirilla are nitrogen fixing bacteria with nitrogenase activity comparable to those of other nitrogen fixers. The bacteria are microaerophilic actively moving towards self-created gradients of oxygen concentration. Studies with inoculated plants in the field have shown that Azospirillum comprises 1-10% of the total bacterial numbers in the rhizosphere. It has been demonstrated that Azospirillum inoculated plants take up minerals (N, P and K) from solution at faster rates than controls and consequently inoculated plants in the field accumulate more dry matter, N, P and K in their stems and leaves. Under water stress conditions inoculated plant show improvement in the water content of the plants. Inoculated plants also recorded more highly branched and developed root system. From the above observations it is clear that Azospirillum inoculation benefits plant growth and increases yields of crops by improving root development, mineral uptake and plant water relationships. The effect of Azospirillum inoculation on the total yield increase of field grown crops generally ranged from 10 to 30 per cent. In addition to nitrogen fixation Azospirillum also produces growth promoting substances like IAA, GA, Kinetin, *etc.* and these phytohormones go a long way in enhancing the crop growth.

Azotobacter

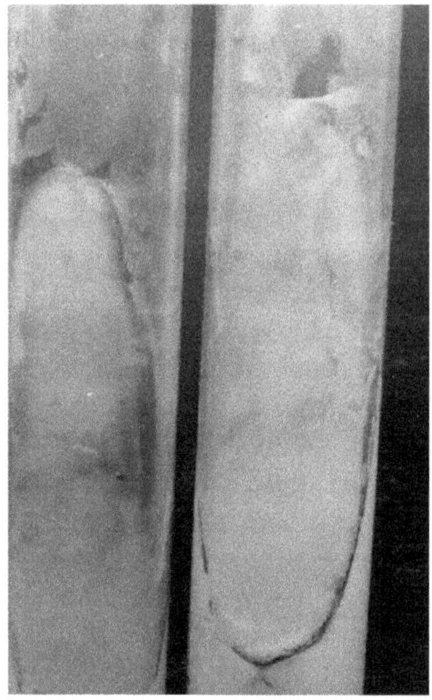

These are free-living bacteria, which grow well on a nitrogen free medium. These bacteria utilize atmospheric nitrogen gas for their cell protein synthesis. This cell protein is then mineralized in soil after the death of Azotobacter cells thereby contributing towards the nitrogen availability of the crop plants. Azotobacter is Gram-negative bacteria, polymorphic *i.e.* they are of different sizes and shapes. Their size ranges from 2-10x1-2.5 m, young cell possess peritrichous flegella and are used as locomotive organs. Old population of bacteria includes encapsulated forms and have enhanced resistant to heat, desiccation and adverse conditions. The cyst germinates under favourable conditions to give vegetative cells. They also produce polysaccharides. Azotobacter spp., are sensitive to acidic pH, high salts, and temperature above 350C. There are four important species of Azotobacter *viz.* A. chroococcum, A. agilis, A.paspali and A. vinelandii of which A. chroococcum is most commonly found in our soils.

The species of Azotobacter are known to fix on an average 10 mg. of N/g of sugar in pure culture on a nitrogen free medium. A maximum of 30 mg. N fixed per gram of sugar was reported by lopatina. However, Azotobacter is a poor competitor for nutrients in soil. Most efficient strains of Azotobacter would need to oxidize about 1000 kg of organic matter for fixing 30 kg of N/ha. This does not sound realistic for our soils, which have very low active carbon status. Besides, soil is inhabitated by a large variety of other microbes, all of which compete for the active carbon.

In Indian soils, the population of Azotobacter is not more than 10 thousand to 1 lakh/g of soil. Other microorganisms present in soil mostly influence the Population of Azotobacter. There are some microorganisms, which stimulate the Azotobacter population in soil thereby increasing the nitrogen fixation by Azotobacter. On the other hand there are some microorganisms, which adversely affect the Azotobacter population, and hence nitrogen fixation process is hampered. For example Cephalosporium is most commonly found organisms in soil, which restricts the growth of Azotobacter. Azotobacter also produces some substances, which check the plant pathogens such as Alternaria, Fusarium and Helminthosporium. Hence

Azotobacter also acts as a biological control agent.

Azotobacter naturally fixes atmospheric nitrogen in the rhizosphere. There are different strains of Azotobacter each has varied chemical, biological and other characters. However, some strains have higher nitrogen fixing ability than others.. Azotobacter uses carbon for its metabolism from simple or compound substances of carbonaceous in nature. Besides carbon, Azotobacter also requires calcium for nitrogen fixation. Similarly, a medium used for growth of Azotobacter is required to have presence of organic nitrogen, micro-nutrients and salt in order to enhance the nitrogen fixing ability of Azotobacter.

Besides, nitrogen fixation, Azotobacter also produces, Thiamin, Riboflavin, Nicotine, indol acetic acid and gibberellins. When Azotobacter is applied to seeds, seed germination is improved to a considerable extent, so also it controls plant diseases due to above substances produced by Azotobacter.

Gluconacetobacter

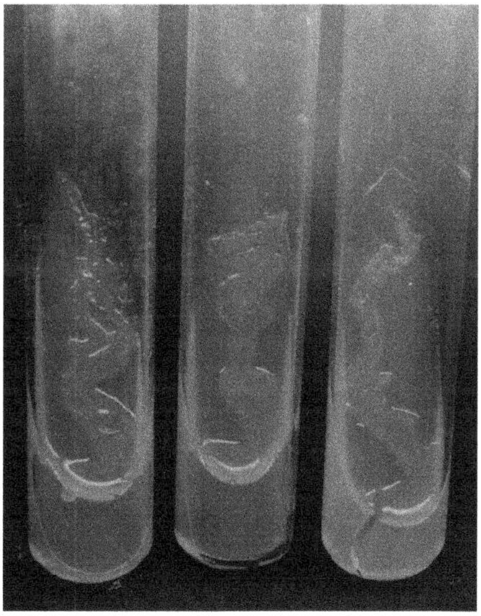

Although a number of microorganisms belonging to different genera have been identified to be associated with sugarcane and fix atmospheric nitrogen, only microorganisms, which belong to the genera Azospirillum and Azotobacter, have been considered to be of economical importance. In recent years, explanation for large scale contribution of biological nitrogen fixation in crops like sugarcane and non-legumes has become much easier by description of newer obligate endophytic bacterium Gluconacetobacter diazotrophicus which get colonized within roots, stems and leaves of certain plant genotypes. In addition to Gluconacetobacter diazotrophicus, Herbaspirillum seropedicae Herbaspirillum rubrisubalbicans and Azoarcus have also been reported as other endophytic bacteria. However physiological and nutritional studies with sugar rich crops *viz.*, sugarcane, sweet potato, sweet sorghum and coffee plants revealed the importance of nitrogen fixing bacterium, Gluconacetobacter diazotrophicus. In addition to nitrogen fixation, these organisms also produce Indole-3-acetic acid, which serve as a growth promoter.

Salient features

- ❑ Acid tolerant nitrogen fixing bacterium associated with sugarcane.

- An endophytic organism colonizes the roots, stems, buds and leaves of sugarcane
- It is located in the apoplastic fluid of sugarcane stems and to a lesser extent in xylem vessels.
- Capable of growing and fixing nitrogen at 30% sucrose concentration and optimum sucrose concentration for better nitrogen fixation is 10%
- Growth and Nitrogen fixation observed even at pH 3. 0
- The optimum pH is 5.5
- An average of 200 kg N ha^{-1} would be fixed by using 30 per cent of the total sugar produced by a sugarcane crop.
- IAA production is also noticed
- Tolerant to higher level of combined nitrogen.
- Mode of entry is through VA mycorrhizae, mealy bug and mechanical injury of plant parts
- Exhibiting antibiotic resistance
- It is a member of alpha group of Proteobacteria
- In terms of 16s rDNA based phylogeny Gluconacetobacter clusters with the genera Rhodopila, Acidomonas, Acidiphilium and Gluconobacter
- In terms of nif H sequence, it is more related to Rhizobium sp. and Azospirillum brasilense than to other diazotrophs.

Phosphobacteria

It has long been known that the microorganisms play an important role in the mobilization and immobilization of phosphorus. Phosphate solubilising microorganism distributed well in several ecosystems, type and nature of microorganisms depend on the type of soil and the vegetation that comes on it. Various microorganisms such as bacteria, actinomycetes, cyanobacteria, fungi and yeast are known to solubilize insoluble inorganic as well as bound organic phosphates.

Although phosphorus is one of the most important macronutrients of all living organisms, many processes of the P cycle are still obscure. Several P-forms might be taken up by the cell, but by far the greatest part is absorbed in the forms of HPO_{42}^{-} or H_2PO_4. Therefore free inorganic phosphates in soil solution play a central role in the cycling of P. Besides fertilization thin fraction is mainly increased by two different pathways.

The enzymatic decomposition of organic P compounds : *e.g.* phosphomonoesters, phosphodiesters, including phospholipids and nucleicacids, phosphotriesters *etc.* and inorganic P compounds.

The non enzymatic solubilization of different rock phosphates and inorganic phosphorus source.

A large fraction of soil microbial population can dissolve insoluble inorganic phosphates present in soils. In some soils this ability has been found to be present in as high as 8% of the total population. In addition to bacteria, actinomycetes belonging to the genera Micromonospora, Nocardia and Streptomyces have also been reported to solubilize insoluble phosphates. Thermophilic fungi which are normally not present in soil have been isolated from compost and characterised for solubilization of tricalcium phosphate and rock phosphate. Torula thermophila solubilized Mussoorie rock phosphate and utilized it as a sole P source efficiently.

The soil environment surrounding plant roots is the zone of intense microbial activity. A large number of microorganisms capable of solubilising insoluble phosphates have been isolated from the root region of crop plants. The population of phosphate dissolving microorganisms is more in the rhizosphere (20-40% of the total population) compared to non-rhizosphere (10-15% of the total population). A large number of anaerobic and aerobic phosphate dissolving bacteria were found in these soils. Both anaerobic and aerobic phosphate dissolving bacteria were stimulated in the rice rhizosphere, but the rhizosphere effect was greater in case of anaerobic phosphate dissolving bacteria. The higher population of phosphate solubilizers in the rhizosphere is of great relevance to plants especially in P deficient soils as it helps in mobilization of insoluble P.

Phosphate solubilising bacteria

Achromobacter sp

Aerobacter sp.

Agrobacterium sp

Bacillus sp.

B. megaterium var. *Phosphaticum*

Escherichia sp

Pseudomonas sp

Serratia sp.

Phosphate solubilizing cyanobacteria

Anabaena sp

Calothrix

Nostoc sp

Scytonema sp

Tolpothrix

Phosphate solubilizing actinomycetes

Actinomyces sp.

Streptomyces

Phosphate solubilizing fungi and yeast

Aspergillus spp.

A. awamorii

Mucor sp

Penicillium sp

Pythium sp.

Fusarium sp

Rhizoctonia solani

Helminthosporium sp

Organic acids and P solubilization

The major microbiological means by which insoluble phosphorus compounds are mobilized is by the production of organic acid which is accompanied by the acidification of the medium. The organic and inorganic acids convert tricalcium phosphates to the di- and monobasic phosphates with the net result of an enhanced availability of the element to the plant. The type of organic acid produced and their amounts differ with different microorganisms. Tri- and di-carboxylic acids are more effective as compared to mono basic and aromatic acids. Aliphatic acids are also found to be more effective in P solubilization than phenolic acids and citric acid and fumaric acids has highest P-solubilizing ability.

Enzymes and P-solubilization

The liberation of P from organic phosphate compounds is mainly due to the action of enzymes of esterase type. PSM along with acid production, produces the phosphatase enzyme which cause the solubilization of P in aquatic environment. Besides these two mechanisms the production of chelating substances, H_2S, CO_2, mineral acids, siderophores are also involved in P solubilization by PSMs. In addition, the action of PSMs is not only due to the release of plant available phosphorus but also due to the production of biologically active substances like indole acetic acid, gibberellins, cytokinins production is also correlated with P solubilization.

Mycorrhiza

Mycorrhizal symbiosis is the most prevalent and important mutualistic association in plant kingdom. Ectomycorrhizal fungi are limited to mostly temperate forest trees and endomycorrhizae form external hyphal net work in soil and grow extensively with in the cells of the cortex, found all most in all crop plants. Members of Ericaceae (Ericoid Mycorrhizae) find specific types of Endomycorrhizae. AM fungi are the most wide spread type of endomycorrhizae and considered as obligate symbiont to crop plants for better utilization of phosphorus and other essential elements.

Ectomycorrhizal Fungi

Among the different mycorrhizas, the ectomycorrhizal fungi occurring in about 10% of the world's flora include most of the gymnosperms especially the conifers and some of the angiosperm species like *Eucalyptus, Casuarina,Acacia, Betula.* The following ectomycorrhizal fungi were found in the different plantations :*Amanita muscaria, Laccaria laccata, Lycoperdon perlatum, Rhizopogon luteolus, Russula* sp., *Suillus brevipes, S. subluteus, Scleroderma citrinum* and *Thelephora terrestris.*

Ectomycorrhiza

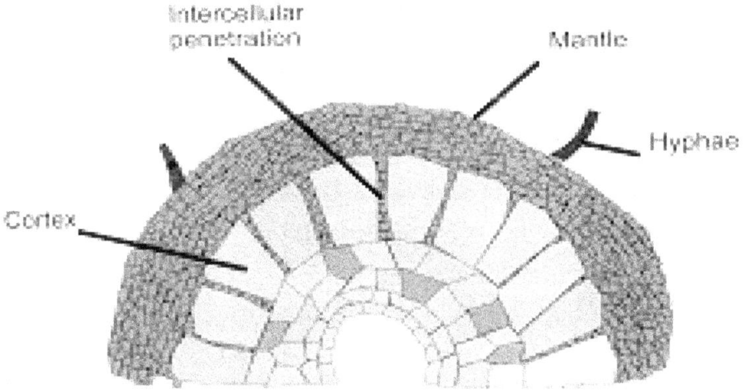

– Endomychorhizal fungi – solubilization and mobilization of phosphorus – mechanisms involved.

Ectomycorrhizas (sometimes termed ectotrophic mycorrhizas) are characteristic of many trees in the cooler parts of the world - for example pines, spruces, firs, oaks, birches in the Northern Hemisphere and eucalypts in Australia. However, some trees (*e.g.* willows) can have both ectomycorrhizas and arbuscular mycorrhizas, and most tropical trees have only arbuscular mycorrhizas.

The fungi involved are mainly Ascomycota and Basidiomycota, including many that produce the characteristic toadstools of the forest floor. Most of these fungi can be grown in laboratory culture but, unlike the wood-rotting fungi, they are poor degraders of cellulose and other plant wall materials. So they gain most of their sugars from the living plant roots in natural conditions.

Ectomycorrhizal fungi usually form a mantle that encloses the plant rootlet, from which hyphae (*i.e.*,vegetative, non-reproductive, threadlike filaments) radiate outward into the soil, as well as inward between the root cells to form a hyphal network called the"Hartig net." The hyphae of most ectomycorrhizal fungal species proliferate in the duff layer of the forest floor, but some also inhabit mineral soil, and still others prefer decaying wood as a substrate. Some ectomycorrhizal fungi require large amounts of carbohydrate, which they acquire from their plant hosts, and so are dependent on mature trees that can meet their carbohydrate demands. Although all ectomycorrhizal fungi produce spores, only some of these species seem able to colonize from them. Nevertheless, spores may be especially important after fire that is severe enough to burn organic soil horizons. Spores of mushrooms are dispersed by wind and thus can reach the centre of burns and clear cuts. Spores of truffles are dispersed primarily by small mammals. Once trees are harvested, the ectomycorrhizae remaining in the soil appear to take 1–2 years to die. Dying ectomycorrhizal roots sometimes support sufficient live hyphae to colonize new roots, depending on the fungal species involved. Thus, they are a possible source of inoculum following clear cut harvesting. Sclerotia are produced by relatively few ectomycorhizal fungal species (*e.g.*, *Cenococcum geophilum*). Sclerotia can persist from a few weeks to months, depending on the fungal species, and can be especially abundant following fire. Sclerotia contain sufficient hyphae to act as inoculum.

Ectomycorrhizal Fungal Diversity

Ectomycorrhizal fungal species differ in several ways, including :

- Their ability to take up various forms and types of nutrients,
- Their rate of nutrient uptake,
- Their tolerance to water stress and temperature extremes,
- The substrates they inhabit, and
- The parts of the root system (distance from the bole) with which they form the associations. Because of the variability in the characteristics of ectomycorrhizal fungi, trees forming a diverse array of ectomycorrhizae are thought to be better suited to survive and grow in variable soil and climatic conditions than trees forming ectomycorrhizae with only one or a few fungal species

The Functional Role of Ectomycorrhizal Fungi

Ectomycorrhizal fungi play an important role in tree growth. They provide numerous benefits to their host plant with different fungal species providing different benefits. These include:

- Enhancing the uptake of essential nutrients (mainly phosphorus and nitrogen) and water;
- Protecting against pathogens and heavy metals;
- Binding soil particles to create favourable soil structure;
- Facilitating below-ground nutrient transfer among plants and
- Altering the competitive relationships among plants of different species.

One type of ectomycorrhiza, known as a tuberculate ectomycorrhiza, also harbours nitrogen-fixing bacteria.

Functioning of ectomycorrhizas

Experiments with radioactive tracers have shown that when labelled CO_2 is applied to leaves of tree seedlings, the label is found in plant sugars (sucrose, *etc.*) which move to the roots; then the label enters the fungal sheath where it occurs in the form of typical 'fungal carbohydrates' such as mannitol and trehalose. Most plants and plant tissues cannot metabolise these compounds, so there is, in effect, a one-way flow of carbohydrate to the fungus. The cost of this to the plant may be considerable. However, the plant also can benefit from the association, because the fungal hyphae that ramify into soil are very efficient in capturing mineral nutrients; these accumulate in the sheath but at least some minerals are transferred to the plant, presumably from the Hartig net.

Recent work has shown that several ectomycorrhizal fungi can degrade proteins (they release protease enzymes) and thus can obtain nitrogen from the decomposing leaf litter. This could be highly significant in temperate and sub-boreal forests, where the rates of mineral nutrient recycling are low because of the low microbial activity in cool, acidic conditions. Ectomycorrhizal fungi might thus play a key role in the nitrogen nutrition of trees. It has also been shown that young seedlings growing in the shade of 'mother' trees can be attached to the 'parent' by a common network of mycorrhizal hyphae in the soil. In these conditions, at least some movement of labelled carbohydrates has been shown to occur from theroots of the parent to the younger trees, perhaps helping to nurture the seedlings Ectomycorrhizae and plant disease.

A pathogen attacking a ectomycorrhizal root system is initially confronted externally with the tightly interwoven network of hyphae that makes up the fungal mantle and then internally with cortical cells whose cell walls are surrounded by Hartig net hyphal tissue. A pathogen of a mycorrhizal root system must have the physical and chemical ability to penetrate these barriers. Mycorrhizal tree seedlings have the ability to resist root infections by potential pathogens.

Mechanisms of disease protection by ectomycorrhizae

a) Utilization of surface carbohydrates in roots by mycorrhizal endophytes eventually reducing the quantum of nutrients available to pathogens.

b) Formation of mechanical barrier in the form of fungal mantle aganint ingress by the pathogen

c) Secretion of antibiotics inhibitory to the pathogens

d) Formation of a protective microbial rhizospphere population

e) The production of metabolites released by mycorrhizal host cortical cells functioning as inhibitor of infection and spread of the pathogen.

The fungal mantle of ectomycorrhizal roots form formidable physical barriers that restrict penetration by several root pathogens particularly *Phytophthora cinnamomi*. The hartig net surrounding the cortical cells may function as a supplementary physical barrier since the progress of the pathogen is blocked in this region. The biochemical and cytological response of the cortical cells in the ectomycorrhizal root systems may have a role in disease protection. The fungal endophyte *Suillus variegatus* liberates fungistatic volatile compounds isobutanol, isobutryic acid in pure culture. Volatile organic compounds have been extracted from intact seedling roots of *Pinus sylvestris* grown in pure culture with and without the fungal symbiont. Increased production and accumulation of volatile terpenes and sesquiterpenes was seen in mycorrhizal roots systems compared with non mycorrhizal root systems. Several of these compounds are fungistatic, synthesized as a non specific response of the host cells to symbiont infection. Vapours of volatile compounds like 3- carene and 5 phellandrene. Mycorrhizal strain of Laccaria laccata which can protect primary roots of Douglas fir seedlings from root rot infection by *Fusarium oxysporum* induced an accumulation of phenolic osmophilic material in cortical cells of primary roots. Strains of mycorrhizal fungi which colonize the primary root rapidly compete with other soil organisms aggressively and induce a consistent deposition of phenolics in roots may provide biological control. Root protection by mycorrhizal fungi may be effective not only inside the root but also in the rhizosphere and rhizoplane. Disease severity and plant mortality were shown to be reduced in the presence fungi even before the mycorrhizal formation took place.

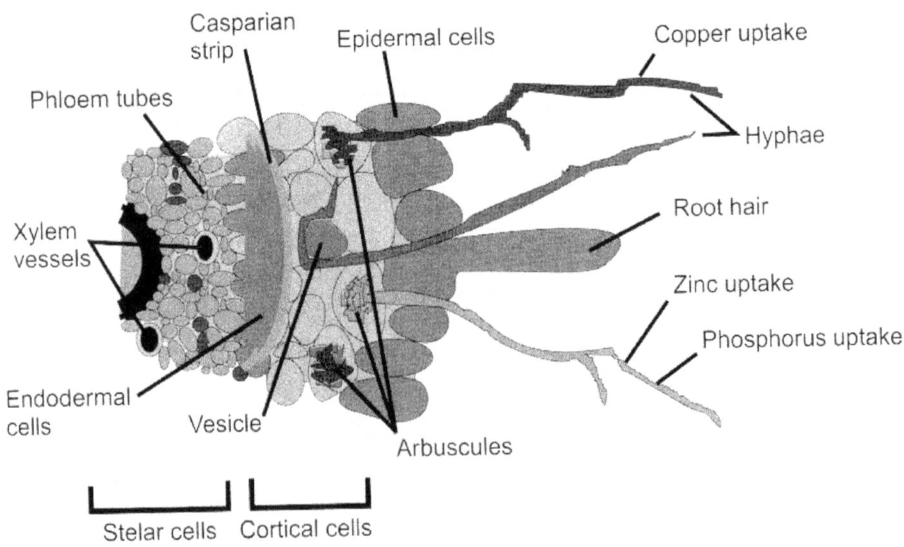

Endomycorrhizae

AM symbiosis is often associated with improved plant growth. This enhanced growth has been attributed to nutritional and non-nutritional effects of AM fungi. It has been reported to benefit plants by increasing the uptake of nutrients such as P, Zn, Cu, and nitrogen. The non nutritional effects of AM would be due to increased tolerance to saline conditions, improved water relations increased survival ratio of transplanted seedlings, control of root diseases and increased soil aggregation by the external hyphal net work. The Mycorrhizal fungi are also known to produce wide array of plant growth promoting substances (IAA, IBA, GA).

Arbuscular Mycorrhizae

Roots of most crop plants live in association with certain beneficial fungi, the association being called mycorrhiza. Vesicular arbuscular mycorrhiza is by far the most wide spread type and about 80% of the land plants including agronomically important crops from this type of mycorrhiza.

AM associations involve fungi in the Zygomycete order Glomales and the roots of a wide diversity of plants. About 150 sp in the six genus *viz. Acaulaspora, Entrophosphora, Gigaspora, Glomus, Sclerocystis* and *Scutellospora* are able to form VAM. None of these fungi has yet been successfully cultured axenically.

Components of VAM associations are:

Structures in Roots

Hyphae - these are non-septate when young and ramify within the cortex.

Arbuscules - intricately branched haustoria in cortex cells.

Vesicles - storage structures formed by many fungi.

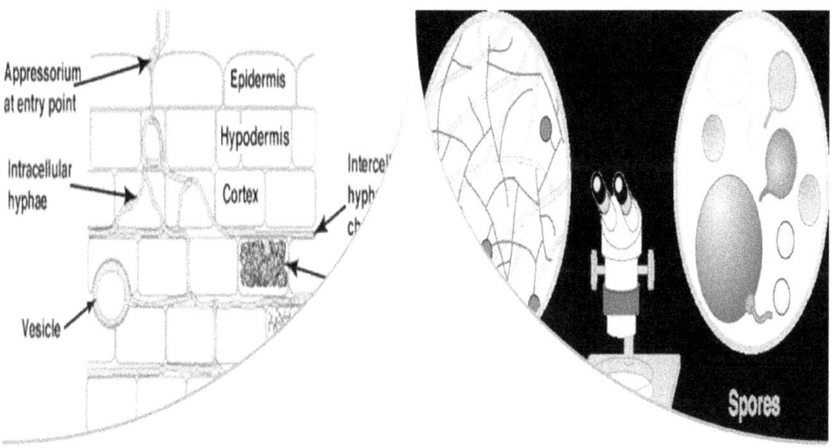

Structures in Soil

Hyphae - A network of hyphae forms in the soil with thicker hyphae which function as conduits and thin branched hyphae which are thought to absorb nutrients.

Spores - Large (for a fungus) asexual spherical structures (20-1000+ um diameter) that form on hyphae in soil or roots.

Structures and Developmental Stages

VAM associations form when host roots and compatible fungi are both active in close proximity and soil conditions are favourable

Soil Hyphae

Soil hyphae, also known as extraradical or external hyphae, are filamentous fungal structures which ramify through the soil. They are responsible for nutrient acquisition, propagation of the association, spore formation, *etc*. VAM fungi produce different types of soil hyphae including thick "runner" or "distributive" hyphae as well as thin "absorptive" hyphae. The finer hyphae can produce "branched absorptive structures" (BAS) where fine hyphae proliferate. Hyphae of VAM fungi have been observed to proliferate in soil microsites where nutrients are concentrated. Hyphae of Scutellospora and Gigaspora species produce clustered swellings with spines or knobs called auxiliary cells (bodies or vesicles).

Root Contact and Penetration

Mycorrhizal associations start when soil hyphae respond to the presence of a root by growing towards it, establishing contact and growing along its surface. Next, one or more hyphae produce swellings called appressoria between epidermal cells. Root penetration occurs when hyphae from the appressoria penetrate epidermal or cortical cells to enter the root. These hyphae cross the hypodermis (through passage cells if these are present in an exodermis) and start branching in the outer cortex

Hyphal Proliferation in the Cortex

Aseptate hyphae spread along the cortex in both directions from the entry point to form a colony. Hyphae within root are initially without cross walls, but these may occur in older roots. VAM associations in different species formed two distinctive morphology types, called the Arum and Paris series after host plants. Both Arum and Paris type VAM associations are important in ecosystems.

In roots with Arum series associations, VAM hyphae proliferate in the cortex by growing longitudinally between host cells. This occurs because hyphae grow through longitudinal intercellular air spaces that are present. These are associations where hyphae grow along longitudinal intercellular air channels between the walls of root cells. A relatively rapid parallel spread of intercellular hyphae may occur along these channels. The resulting colonies of VAM fungi have a linear appearance.

In Paris series VAM hyphae spread by forming coils within cells because there are no continuous longitudinal air spaces. These are associations where hyphae spread primarily by intracellular growth following a convoluted path through cortex cells. The resulting colonies of VAM fungi generally have a coiled appearance but may have more digitate branching patterns. Arbuscules may be restricted to a single layer of cells in the inner cortex.

Arbuscules

Arbuscules are intricately branched haustoria that formed within a root cortex cell. Arbuscules are formed by repeated dichotomous branching and reductions in hyphal width, starting from an initial trunk hypha (5-10 um in diameter) and ending in a proliferation of fine branch hyphae (< 1 um diameter). Arbuscules start to form approximately 2 days after root penetration. They grow inside individual cells of the root cortex, but remain outside their cytoplasm, due to invagination of the plasma membrane. Arbuscules are considered the major site of exchange between the fungus and host.

This assumption is based on the large surface area of the arbuscular interface, but has not been confirmed. Arbuscule formation follows hyphal growth, progressing outwards from the entry point. Arbuscules are short-loved and begin to collapse after a few days, but hyphae and vesicles can remain in roots for months or years.

Vesicles

Vesicles develop to accumulate storage products in many VAM associations. Vesicles are initiated soon after the first arbuscules, but continue to develop when the arbuscules senesce. Vesicles are hyphal swellings in the root cortex that contain lipids and cytoplasm. These may be inter- or intracellular. Vesicles can develop thick walls in older roots and may function as propagules. Some fungi produce vesicles which are similar in structure to the spores they produced in soil, but in other cases they are different.

The most important feature common to mycorrhizal fungi are their ability to recognize the roots of appropriate host plants to infect root tissues without eliciting rapid or intense host defense reactions. The ability of mycorrhizal fingi to contribute to host nutrition is based on soil borne beneficial fungi and therefore their status as soil borne beneficial fungi and therefoe their capacity to gain access to soil derived nutrients. The mycelium produced by mycorrhizal fungi can

1) Survive as resting propagule in the absence of host
2) Grow independently in soil
3) Grow as symbiotic mycelium in soil
4) Grow in close association with roots
4) Form reproductive organs. These fungi in addition to supplying p to the plant offer protection against root diseases. They also help in water uptake and overcoming transplant shock thus help in better establishment of plants.

P contribution

Phosphate transport by VAM is a key factor in plant nutrition. VAM can absorb several times soil P and have greater P inflow rates than roots. The fungal hyphae actually transport P over large distances into the root. The phosphate absorbed by VAM fungi from soil solution is accumulated in the vacuoles of the fungi as poly phosphate granules and this represents 16-40% of the total P in VAM fungi. The poly beta granules are propelled through the hyphae by cytoplasmic streaming to the arbuscules. The inflow rate through external hyphae is approximately thousand times faster than the diffusion rate through the soil pores. It is also important that when phosphate supply is in excess polyphosphate accumulates the nutrient as a storage pool that can be used at times of increasing demand to support stages of plant growth. Increased uptake of P due to inoculation with VAM fungi in various agricultural crops has been reported.

The ability of mycorrhizal fungi to contribute to host nutrition is based on their status as soil-borne, beneficial fungi and therefore their capacity to gain access to soil derived nutrients. The mycelium produced by mycorrhizal fungi can:

- Survive as resting propagule in the absence of host
- Grow independently in soil
- Grow as symbiotic mycelium in soil (extra hyphal structure)
- Grow in close association with the roots (intracelluar hypae and arbuscles in VAM fungi)

Form reproductive or resting structures

These fungi in addition to supplying P to the plant offer protection against root diseases. They also help in water uptake and overcoming transplant shock, thus help in better est

Azolla

Azolla is a green - floating water fern that floats in water and fixes atmospheric nitrogen in association with nitrogen fixing blue green alga *Anabaena azollae* (Moore, 1969). *Azolla* fronds consists of sporophyte with a floating rhizome and small overlapping bilobed leaves and roots. Rice growing areas in South East Asia and other third World countries have recently been evincing increased interest in the use of the symbiotic N_2 fixing water fern *Azolla* either as an alternate nitrogen source or as a supplement to commercial nitrogen fertilizers.

Azolla is used as biofertilizer for wetland rice and it is known to contribute 40-60 kg N ha^{-1} per rice crop. The agronomic potential of *Azolla* is quite significant particularly for rice crop and it is widely used as biofertilizer for increasing rice yield.. The important factor in using *Azolla* as a biofertilizer for rice crop is its quick decomposition in soil and efficient availability of its nitrogen to rice. In tropical rice soil, the applied *Azolla* mineralizes rapidly and its nitrogen is made available to the rice crop in very short period.

Azolla -Anabaena symbiosis

The eukaryotic water fern *Azolla* and the prokaryotic endosymbiont*Anabaena azollae* are actively involved in the symbiotic association and fix atmospheric nitrogen. The host plant *Azolla* provides the carbon source sucrose to the cyanobiont while the cyanobiont fix nitrogen and transfers to *Azolla* which in turn multiplies very rapidly. The cyanobiont is associated with dorsal lobes of the leaf from the onset of

their development and is never in direct contact with the external environment and *A. azollae* is the only species found in symbiotic association with *Azolla*. The growth and development of the *Azolla* and the *Anabaena* are synchronous. *Azolla* shoot tips are supported above the water surface by the stem axes and a small colony of *Anabaena* filaments is associated with every apical merits. The heterocyst of the symbiont *Anabaena* is the site of nitrogen fixation, while vegetative cells are involved in photosynthesis. The total nitrogen requirement of the *Azolla* fronds is supplied by the cyanobiont.

The ammonia from N_2 fixation by the cyanobiont is released into the leaf cavity and absorbed by the frond's ammonia assimilating enzymes. The growth rate of *Azolla* is very rapid so much so that it doubles its weight in a mere 2-3 days. Due to fast relative growth rate the biomass production is heavy in a given time and hence the biomass is exploited as biofertilizer for rice. *Azolla* has the potential of maintaining an exponential growth rate under optimum conditions.

Blue Green Algae

Blue-green algae otherwise called as cyanobacteria are oxygen evolving and nitrogen fixing prokaryotes, using sunlight as the sole energy source for the fixation of carbon and nitrogen. Many cyanobacteria also contribute greatly to the nitrogen economy of aquatic and terrestrial habitats through their ability to fix atmospheric nitrogen. Besides fixing atmospheric nitrogen, cyanobacteria play a major role in reducing soil erosion, increasing the soil organic matter content and in producing certain substances which enhance the growth of higher plants. Algal application resulted in a significant improvement in the aggregation status of the soil and brought down appreciably the pH, EC, CEC and increased the hydraulic conductivity of the soil.

In recent years, the practice of utilizing cyanobacteria as an efficient source of biofertilizer for rice have been advocated and adopted in India. The rice field ecosystem provides the conductive environment for them with respect to light, water, high temperature and nutrient availability. Amongst the various soil factors affecting the distribution of cyanobacteria, pH is an important one which directly influence their distribution and abundance in soil. Cyanobacteria generally prefer neutral to slightly alkaline condition. Many exhibit salt tolerance due to the presence of extracellular polysaccharides. The salt tolerance of cyanobacteria has been exploited with some success in reclamation of saline and sodic soils. Application of cyanobacteria to saline soil enhances the soil organic carbon content due to the addition of enormous quantity of organic matter in the form of algal biomass and reduces the EC of saline soil. The salt tolerant cyanobacteria immobilize Na^+ in an organic form and thereby play a major role in saline soil.

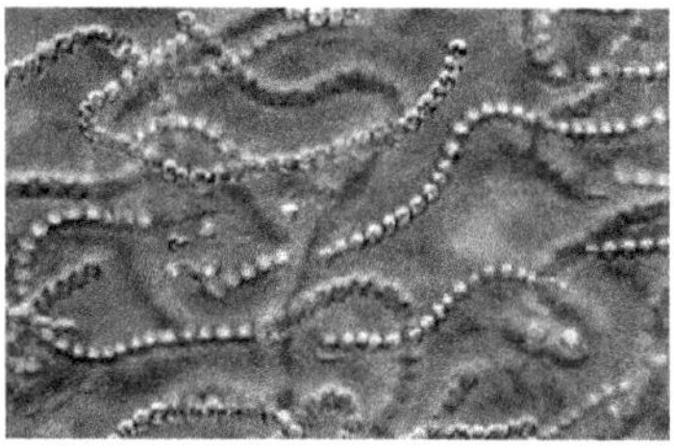

The species of cyanobacteria which are known to fix atmospheric nitrogen are classified into three groups (i) Heterocystous – aerobic forms, (ii) Aerobic unicellular forms and (iii) Non heterocystous, filamentous, microaerophilic forms. The cyanobacterial cultures composite inoculum consisting of *Nostoc, Anabaena, Calothrix, Tolypothrix, Plectonema, Aphanothece, Gloeocapasa, Oscillatoria, Cylindrospermum, Aulosira* and *Scytonema* have been used for inoculation in rice. These N_2 fixing cultures have better adaptability in wet land rice.

Algal population appears to be highly susceptible to environmental changes and exhibit rapid qualitative and quantitative variation along the cultivation cycle. Development of dense algal blooms just after transplanting due to fertilizers or ploughing or both and a high light availability have been reported. *Anabaena* was the first species to develop two weeks after transplanting. N_2 fixing forms *viz.Scytonema, Aulosira, Nostoc, Aphantheca, Oscillatoria, Plectonema* were dominant during tillering phase of rice crop. During the cultivation cycle a decrease in light intensity and N level related to rice growth and an increase in pH value favoured cyanobacterial growth. *Gloeotrichia* and *Rivularia* are reported to occur during panicle initiation and flowering phase of rice.

Silicate and Zinc solubilizing bacteria (SSB)

Microorganisms are capable of degrading silicates and aluminum silicates. During the metabolism of microbes several organic acids are produced and these have a dual role in silicate weathering. They supply H+ ions to the medium and promote hydrolysis and the organic acids like citric, oxalic acid, Keto acids and hydroxy carbolic acids which from complexes with cations promote their removal and retention in the medium in a dissolved state. The studies conducted with a Bacillus sp. isolated from the soil of granite crusher yard showed that the bacterium is capable of dissolving several silicate minerals under in vitro condition. The examination of anthrpogenic materials like cement, agro inputs like super phosphate and rock phosphate exhibited silicate solubilizing bacteria to a varying degree. The bacterial isolates made from different locations had varying degree of silicate solubilizing potential. Soil inoculation studies with selected isolate with red soil, clay soil, sand and hilly soil showed that the organisms multiplied in all types of soil and released more of silica and the available silica increased in soil and water. Rice responded well to application of organic sliceous residue like rice straw, rice husk and black ash @ 5 t/ha. Combining SSB with these residues further resulted in increased plant growth and grain yield. This enhancement is due to increased dissolution of silica and nutrients from the soil.

Biofertilizers Production

India produces about 12,000 tonnes of biofertilizers like *Rhizobium, Azospirillum,* Azotobacter, Acetobacter, Phosphate Solubilizing Microorganisms (PSM) per annum by Central Govt., State Govts., Public Sectors, Autonomous Govt. bodies, Fertilizer Factories as well as through small Private entrepreneurs whose number are large as compared to few large scale producers.

Strain Selection and Improvement

In biofertilizer technology, strain selection occupies the foremost key area of technology, so as to explore its biological potentialities of the strain for the beneficiary crop. The selected strain used in biofertilizer production must be efficient to grow fast besides retain desirable potentialities of biological activities, which should be present in desired high no. and must be stable.

Biofertilizers are defined as preparations containing living cells or latent cells of efficient strains of microorganisms that help crop plants' uptake of nutrients by their interactions in the rhizosphere when applied through seed or soil. They accelerate certain microbial processes in the soil which augment the extent of availability of nutrients in a form easily assimilated by plants.Several microorganisms and their associations with crop plants are being exploited in the production of biofertilizers. N_2 fixing organism such as *Azospirillum, Rhizobium, Azotobacter, Gluconacetobacter* and PO_4 solubilizing bacterial genera *Pseudomonas, Bacillus* and PO_4 mobilizing Arbuscular mycorrhizal fungi are presently used as biofertilizers for commercial application.

I. Mass Production of bacterial biofertilizers

Biofertilizers are carrier based preparations containing efficient strain of nitrogen fixing or phosphate solubilizing microorganisms. Biofertilizers are formulated usually as carrier based inoculants. The organic carrier materials are more effective for the preparation of bacterial inoculants. The solid inoculants carry more number of bacterial cells and support the survival of cells for longer periods of time.

The mass production of carrier based bacterial biofertilizers involves three stages.

1. Mass culturing of microorganisms
2. Processing of carrier material
3. Mixing of broth culture with the carrier and packing

1. Mass culturing of Microorganisms

Although many bacteria can be used beneficially as a biofertilizer the technique of mass production is standardized for *Rhizobium, Azospirillum, Azotobacter* and phosphobacteria.

The media used for mass culturing are as follows:

Rhizobium : Yeast extract mannitol broth.

Azospirillum: Dobereiner's malic acid broth with NH_4Cl (1g per liter)

Azotobacter : Waksmann No.77 broth

Phosphobacteria: Nutrient broth

The broth is prepared in flasks and inoculum from mother culture is transferred to flasks. The culture is grown under shaking conditions at 30±2°C as submerged culture. The culture is incubated until maximum cell population of 1010 to 1011 cfu/ml is produced. Under optimum conditions this population level could be attained with in 4 to 5 days for Rhizobium; 5 to 7 days for *Azospirillum;* 2 to 3 days for phosphobacteria and 6-7 days for Azotobacter. The culture obtained in the flask is called starter culture. For large scale production of inoculant, inoculum from starter culture is transferred to large flasks/seed tank fermenter and grown until required level of cell count is reached.

Inoculum preparation

- ❑ Prepare appropriate media for specific to the bacterial inoculant in 250 ml, 500 ml, 3 litre and 5 litre conical flasks and sterilize.
- ❑ The media in 250 ml flask is inoculated with efficient bacterial strain under aseptic condition
- ❑ Keep the flask under room temperature in rotary shaker (200 rpm) for 5- 7 days.
- ❑ Observe the flask for growth of the culture and estimate the population, which serves as the starter culture.
- ❑ Using the starter culture (at log phase) inoculate the larger flasks (500 ml, 3 liters and 5 liters) containing the media, after obtaining growth in each flask.
- ❑ The above media is prepared in large quantities in fermenter

Fermenter

Mass culturing requires large capacity fermenters. Fermenter is the vessel which maintains the controlled environmental conditions for the growth of microorganisms and provides access for inoculation, sampling, aeration and cleaning. It should be made of stainless steel to withstand high pressure and also to resist corrosion. High quality fermenter will have smooth surfaces inside.

Basic functions of a fermenter

- The main function of a fermenter is to provide a controlled environment for the growth of microorganisms in order to obtain a desired product.
- Fermenter vessel should be capable of being operated aseptically for a number of days. Should withstand high pressure. The power consumption should be as low as possible. The vessel should be designed to require minimal use of labour for operation.
- The vessel should be constructed to ensure smooth internal surfaces with out cracks and crevices.

Description of a fermenter

The vessel in which industrial fermentation process carried out is called as fermenter. The volume of the fermenter varies from 5-10 lit to 500,000 lit. Similarly, the fermenter model will vary for aerobic process and for anaerobic process. It is domed at the top and bottom and held upright by a stainless steel skirt. The top has accessories like a steam pressure gauge, a safety pressure release valve, aeration system with filters, inoculation port. Inlet and outlet ports are provided for water passage through stainless steel cooling coil. The sampling port is provided at the bottom to withdraw samples or to harvest the cultures. The basic structure of fermenter is as follows:

The components are

Inlet – to feed the substrates, microorganism *etc*

Outlet – to collect the products, to take samples

Stirrer – to mix the contents so as to get contact between nutrients and microbes

Air sparger – to provide oxygen to the microbes by small air bubbles

pH meter/controller – to check the pH and if needed to add acid and alkali to maintain optimum pH

Thermometer – to check the temperature

Outer jacket - by which cool water will be circulated to cool the fermenter (because during fermentation, lot of heat will be generated)

Sterilization unit – by which steam will be supplied to the fermenter to sterilize the medium and container

However, the actual operation volume in a fermenter is always less than that of the total volume (2/3), because it requires head space for splashing, foaming and aeration of the liquid inside.

Sterilization of growth medium in the fermenter

- Prepare required quantity of growth medium and adjust to the required pH
- Pour the medium into the fermenter vessel after closing the sampling valve
- Keep the air outlet valve open

- Bring the growth medium to boiling under maximum heat by using steam generator
- Close the air outlet valve and allow the pressure to build up inside the vessel
- Maintain a pressure of 15 1b / in2 at 121°C for 20 minutes
- Switch off the fermenter and cool the medium by circulating cool water.

Mass culturing in fermenter

- Spray the inoculation port with alcohol and flame thoroughly
- Allow the port to cool, inoculate the media in the fermenter vessel with the log phase culture grown in 5 litre flask. Usually 1 -2 % inoculum is sufficient, however inoculation is done up to 5% depending on the growth of the culture in the larger flasks.
- Turn on the air pump, open the air outlet valve
- Regulate the air flow to 3-10 lit of air per hour per lit. of the medium. The sterile air provides aeration as well as agitation for the growth of culture
- Draw samples and analyze for growth, periodically if necessary
- Once the culture reaches full growth turn off the air supply and the cells are harvested with the population load of 109 cells ml-1 after incubation period through the sampling port.
- There should not be any fungal or any other bacterial contamination at 10-6 dilution level
- It is not advisable to store the broth after fermentation for periods longer than 24 hours. Even at 4o C number of viable cells begins to decrease.

2. Processing of carrier material

The use of ideal carrier material is necessary in the production of good quality biofertilizer. Peat soil, lignite , vermiculite, charcoal,press mud, farmyard manure and soil mixture can be used as carrier materials. The neutralized peat soil/lignite are found to be better carrier materials for biofertilizer production The following points are to be considered in the selection of ideal carrier material.

- Cheaper in cost
- Should be locally available
- High organic matter content
- Non toxic chemicals
- Water holding capacity of more than 50%
- Easy to process, friability and vulnerability.
- Amenable for mixing

Preparation of carrier material:

- The carrier material (peat or lignite) is powdered to a fine powder so as to pass through 212 micron IS sieve.

- The pH of the carrier material is neutralized with the help of calcium carbonate (1:10 ratio) , since the peat soil / lignite are acidic in nature (pH of 4 - 5)
- The neutralized carrier material is sterilized in an autoclave to eliminate the contaminants

3. Mixing of broth culture with the carrier and packing

Inoculant packets are prepared by mixing the broth culture obtained from fermenter with sterile carrier material as described below:

Preparation of Inoculants packet:

- The neutralized, sterilized carrier material is spread in a clean, dry, sterile metallic or plastic tray.
- The bacterial culture drawn from the fermenter is added to the sterilized carrier and mixed well by manual (by wearing sterile gloves) or by mechanical mixer. The culture suspension is to be added to a level of 40 – 50% water holding capacity depending upon the population.
- The inoculant packet of 200 g quantities in polythene bags, sealed with electric sealer and allowed for curing for 2 -3 days at room temperature (curing can be done by spreading the inoculant on a clean floor/polythene sheet/ by keeping in open shallow tubs/ trays with polythene covering for 2 -3 days at room temperature before packaging)

Specification of the polythene bags:

- The polythene bags should be of low density grade.
- The thickness of the bag should be around 50 – 75 micron.
- Each packet should be marked with the name of the manufacturer, name of the product, strain number, the crop to which recommended, method of inoculation, date of manufacture, batch number, date of expiry, price, full address of the manufacturer and storage instructions *etc.*,

Storage of biofertilizer packet:

- The packet should be stored in a cool place away from the heat or direct sunlight.
- The packets may be stored at room temperature or in cold storage conditions in lots in plastic crates or polythene / gunny bags.
- The population of inoculant in the carrier inoculant packet may be determined at 15 days interval. There should be more than 109 cells / g of inoculant at the time of preparation and107 cells/ g on dry weight basis before expiry date.

II. Mass production technology for AM fungi

The commercial utilization of mycorrhizal fungi has become difficult because of the obligate symbiotic nature and difficulty in culturing on laboratory media.

Production of AM inoculum has evolved from the original use of infested field soils to the current practice of using pot culture inoculum derived from the surface disinfected spores of single AM fungus on a host plant grown in sterilized culture medium. Several researches in different parts of the world resulted in different methods of production of AM fungal inoculum as soil based culture as well as carrier based inoculum. Root organ culture and nutrient film technique provide scope for the production of soil less culture.

As a carrier based inoculum, pot culture is widely adopted method for production. The AM inoculum was prepared by using sterilized soil and wide array of host crops were used as host. The sterilization process is a cumbersome one and scientists started using inert materials for production of AM fungi. The researchers tried use of perlite, montmorillonite clay *etc.*, In TNAU vermiculite was tried as substrate for the replacement of soil sterilization, which resulted in the best method of inoculum production.

Method of production

1. Form a trench (1m x 1m x 0.3m) is and line with black polythene sheet to use this as a plant growth tub.
2. Mix 50 kg of vermiculite and 5 kg of sterilized soil and pack in the trench up to a height of 20 cm
3. Spread 1 kg of AM inoculum (mother culture) 2-5 cm below the surface of vermiculite
4. Sow with surface sterilized Maize seeds (5% sodium hypochlorite for 2 minutes)
5. Apply 2 g urea, 2 g super phosphate and 1 g muriate of potash for each trench at the time of sowing seeds. Further apply 10 g of urea is twice on 30 and 45 days after sowing for each trench
6. Test the quality of the inoculum by estimationg the AM colonization in root samples on 30th and 45th day
7. Grow the stock plants for 60 days (8 weeks).
8. Cut all the roots of the plant and mix thoroughly with vermiculite in the trench
9. The inoculum produced consists of a mixture of vermiculite, spores, pieces of hyphae and infected root pieces.
10. Thus within 60 days 55 kg of AM inoculum could be produced from 1 sq meter area. This inoculum will be sufficient to treat 550 m2 nursery area having 11,000 seedlings.

A. Blue green algae

Blue green algal inoculation with composite cultures was found to be more effective than single culture inoculation. A technology for mass scale production of composite culture of blue green algae under rice field condition was developed at TNAU and the soil based BGA inoculum could survive for more than 2 years. At many sites where algal inoculation was used for three to four consecutive cropping seasons, the inoculated algae establish well and the effect persisted over subsequent rice crop. Technologies for utilizing nitrogen fixing organisms in low land rice were the beneficial role of blue green algal inoculation in rice soils of Tamil Nadu.

The blue green algal inoculum may be produced by several methods *viz.*, in tubs, galvanized trays, small pits and also in field conditions. However the large-scale production is advisable under field condition which is easily adopted by farmers.

I. Multiplication in trays

a. Big metallic trays (6′x 3′x 6″lbh) can be used for small scale production
b. Take 10 kg of paddy field soil, dry powder well and spread
c. Fill water to a height of 3″
d. Add 250 g of dried algal flakes (soil based) as inoculum
e. Add 150 g of super phosphate and 30 g of lime and mix well with the soil
f. Sprinkle 25 g carbofuran to control the insects
g. Maintain water level in trays
h. After 10 to 15 days, the blooms of BGA will start floating on the water sources
i. At this stage stop watering and drain. Let the soil to dry completely
j. Collect the dry soil based inoculum as flakes
k. Store in a dry place. By this method 5 to 7 kg of soil based inoculum can be obtained.

II. Multiplication under field condition

a. Select an area of 40 m2 (20m x 2m) near a water source which is directly exposed to sunlight.
b. Make a bund all around the plot to a height of 15 cm and give it a coating with mud to prevent loss of water due to percolation.
c. Plot is well prepared and levelled uniformly and water is allowed to a depth of 5-7.5 cm and left to settle for 12 hrs.
d. Apply 2 kg of super phosphate and 200 g lime to each plot uniformly over the area.
e. The soil based composite starter culture of BGA containing 8-10 species @ 5 Kg / plot is powdered well and broadcasted.
f. Carbofuran @ 200 g is also applied to control soil insects occurring in BGA.
g. Water is let in at periodic intervals so that the height of water level is always maintained at 5 cm.
h. After 15 days of inoculation, the plots are allowed to dry up in the sun and the algal flakes are collected and stored.

Observations

The floating algal flasks are brown green or blue green in colour. From each harvest, 30 to 40 Kg of dry algal flakes are obtained from the plot

Method of inoculation of BGA in rice field

Blue green algae may be applied as soil based inoculum to the rice field following the method described below.

a. Powder the soil based algal flakes very well.

b. Mix it with 10 Kg soil or sand (10kg powdered algal flakes with 10 Kg soil / sand).

c. BGA is to be inoculated on 7-10 days after rice transplanting.

d. Water level at 3-4" is to be maintained at the time of BGA inoculation and then for a month so as to have maximum BGA development

e. A week after BGA inoculation, algal growth can be seen and algal mat will float on the water after 2-3 weeks. The algal mat colour will be green or brown or yellowish green.

Observation

A week after BGA inoculation ,algal growth can be seen and algal mat will float on the water after 2-3 weeks.The algal mat colour will be green or brown or yellowish green.

II. Mass production of Azolla

Azolla is a free-floating water fern that floats in water and fixes atmospheric nitrogen in association with nitrogen fixing blue green alga *Anabaena azollae*. Azolla fronds consist of sporophyte with a floating rhizome and small overlapping bi-lobed leaves and roots. Rice growing areas in South East Asia and other third World countries have recently been evincing increased interest in the use of the symbiotic N_2 fixing water fern Azolla either as an alternate nitrogen sources or as a supplement to commercial nitrogen fertilizers. Azolla is used as biofertilizer for wetland rice and it is known to contribute 40-60 kg N ha-1 per rice crop. The agronomic potential of Azolla is quite significant particularly for rice crop and it is widely used as biofertilizer for increasing rice yields. Rice crop response studies with Azolla biofertilizer in the People's Republic in China and in Vietnam have provided good evidence that Azolla incorporation into the soil as a green manure crop is one of the most effective ways of providing nitrogen source for rice. The utilization of Azolla as dual crop with wetland rice is gaining importance in Philippines, Thailand, Srilanka and India. The important factor in using Azolla as a biofertilizer for rice crop is its quick decomposition in soil and efficient availability of its nitrogen to rice. In tropical rice soils the applied Azolla mineralizes rapidly and its nitrogen is available to the rice crop in very short period. The common species of Azolla are *A. microphylla, A. filiculoides, A. pinnata, A. caroliniana, A. nilotica, A. rubra and A. mexicana.*

I. Mass multiplication of Azolla under field conditions

A simple Azolla nursery method for large scale multiplication of Azolla in the field has been evolved for easy adoption by the farmers

a. Select a wetland field and prepare thoroughly and level uniformly.

b. Mark the field into one cent plots (20 x 2m) by providing suitable bunds and irrigation channels.

c. Maintain water level to a height of 10 cm.

d. Mix 10 kg of cattle dung in 20 litres of water and sprinkle in the field.

e. Apply 100 g super phosphate as basal dose.

f. Inoculate fresh Azolla biomass @ 8 kg to each pot.

g. Apply super phosphate @ 100 g as top dressing fertilizer on 4th and 8th day after Azolla inoculation.

h. Apply carbofuran (furadan) granules @ 100 g/plot on 7th day after Azolla inoculation.

i. Maintain the water level at 10 cm height throughout the growth period of two or three weeks.

II. Method of inoculation of Azolla to rice crop

The Azolla biofertilizer may be applied in two ways for the wetland paddy. In the first method, fresh Azolla biomass is inoculated in the paddy field before transplanting and incorporated as green manure. This method requires huge quantity of fresh Azolla. In the other method, Azolla may be inoculated after transplanting rice and grown as dual culture with rice and incorporated subsequently.

A. Azolla biomass incorporation as green manure for rice crop

- Collect the fresh Azolla biomass from the Azolla nursery plot.
- Prepare the wetland well and maintain water just enough for easy incorporation.
- Apply fresh Azolla biomass (15 t ha-1) to the main field and incorporate the Azolla by using implements or tractor.

B. Azolla inoculation as dual crop for rice

- Select a transplanted rice field.
- Collect fresh Azolla inoculum from Azolla nursery.
- Broadcast the fresh Azolla in the transplanted rice field on 7th day after planting (500 kg / ha).
- Maintain water level at 5-7.5cm.
- Note the growth of Azolla mat four weeks after transplanting and incorporate the Azolla biomass by using implements or tranctor or during inter-cultivation practices.

- A second bloom of Azolla will develop 8 weeks after transplanting which may be incorporated again.
- By the two incorporations, 20-25 tonnes of Azolla can be incorporated in one hectare rice field.

Methods of Application of Different Biofertilizers

The term biofertilizers can be more appropriately called 'Microbial inoculants'. These are defined as preparations containing live or latent cells of efficient strains of nitrogen fixing, phospate solubilizing or cellutolytic microorganisms used for application to seed, soil or composting with the objective of increasing the numbers of such microorganisms and accelerate certain microbial processes to augment the availability of nutrients to plants. The bacteria like *Pseduomonas* and *Bacillus* bring insoluble phospates in soil into slouble forms by secreting organic acids such as acetic, propionic, lactic glycolic, fumaric and succinic acids. Bacterial biofertilizers are supplied as carrier based inoculants. Peat or lignite is used as carrier material. Carrier based bacterial inoculants are applied by the following methods.

1. Seed treatment or seed inoculation
2. Seedling root dip and
3. Main field application

Seed treatment

One packet of the inoculant is mixed with 200 ml of rice gruel to make slurry. The seeds required for an acre are mixed in the slurry so as to have a uniform coating of the inoculant over the seeds and then shade dried for 30 minutes. The shade dried seeds should be sown within 24 hours. One packet of the inoculant (200 g) is sufficient to treat 10 Kg of seeds.

Seedling root dip

This method is used for transplanted crops. Two packets of the inoculant is mixed in 40 litres of water. The root portion of the seedlings required for an acre is dipped in the mixture for 5 to 10 minutes and then transplanted.

Main field application

Four packets of the inoculant is mixed with 20 Kg of dried, powdered farm yard manure and then broadcasted in one acre of main field just before transplanting.

Rhizobium

For all legumes Rhizobium is applied as seed treatment.

Azospirillum / Azotobacter

In the transplanted crops, Azospirillum is inoculated by seed treatment, seedling root dip and soil application methods. For direct sown crops, Azospirillum is applied through seed treatment and soil application.

Phosphobacteria

Inoculation is done by seed treatment, seedling root dip and soil application as is done in case of *Azospirillum*.

Combined application of bacterial biofertilizers

Phosphobacteria can be mixed with *Azospirillum* and *Rhizobium*. The inoculants should be mixed in equal quantities and applied as mentioned above.

Bacterial inoculants should not be mixed with insecticide, fungicide, herbicide and fertilizers. Seed treatment with bacterial inoculant is to be done last when seeds are treated with fungicides. Biofertilizer treated seeds should be sown immediately.

AM fungal inoculant

Nursery application: 100 g inoculum is sufficient for one meter square. The inoculants should be applied at 2-3 cm below the soil at the time of sowing. The seeds/cutting should be sown / planted above the AM inoculum. For polythene bag raised seedlings (Forest trees, coffee & tea *etc*). About 10 g of inoculum is sufficient for each plant raised in poly bags. Mix 10 Kg of inoculum with 1000 Kg of potting mixture and pack the potting mixture in polythene bags before sowing.

AM inoculum required for out planting

Twenty grams of AM inoculum is required per seedling. Apply inoculum at the time of planting.

AM inoculum required for existing trees

Two hundred gram of AM inoculum is required for inoculating one tree. Apply inoculum near the root surface at the time of fertilizer application.

Storage of inoculum

AM bulk inoculum can be stored for a period of six months under room temperature.

Quality control of biofertilizers

Quality of inoculants manufactured is one of the most important factors resulting in their success or failure and acceptance or rejection by the farmers. Basically quality means the presence of the right type of microorganism in active form and desired numbers. Quality has to be controlled at various stages of production, marketing and application.

Quality Control Measures (as per ISI Specifications)

- ❑ Since, Biofertilizers contains live cells, care should be taken during their transportation and storage.
- ❑ They should be kept in a cold place and not exposed to sunlight.
- ❑ Biofertilizers for legumes are crop-specific; therefore, they must be used for the crop for which they are meant.

- Biofertilizers when used under adverse soil conditions, appropriate remedial measures (liming and use of Gypsum) should be followed.
- Biofertilizers must be carrier-based
- Carrier material used should be in form of powder (75-106 micron size.
- It should contain minimum of 10^8 viable cells of microorganisms /gram of the carrier material on dry weight basis.
- It should have a minimum period of six months expiry from date of its
- It should be tree from any contaminant /contamination with other microorganisms.
- pH should be in the range of 6.0-7.5.
- It should induce desired beneficial effects on all those crops, species / cultivars listed on the packet before the expiry date.
- It should be packed in 50-75 micron low density polythene packets

Stages requiring quality control

- During mother culture stage
- During carrier selection and standardisation
- During broth culture stage
- During mixing of broth with carrier
- During packing
- During storage
- Standards for quality testing

Bureau of Indian Standards (BIS) for carrier - based Biofertilizers

BIS have notified specifications for *Azotobacter* (ISI, 1979) and *Rhizobium* (ISI, 1986) long back for use of biofertilizer producers and suggested some modifications. The committee has also suggested specifications for phosphate solubilizing microorganisms and *Azospirillum.*

Indian Standard Specifications for different biofertilizers

Parameters	*Rhizobium*	*Azotobacter*	*Azospirillum*	*PSM*
Base	Carrier based	Carrier based	Carrier based	Carrier based
Cell No. at the time of manufacture	107	106	107	108
Contamination	Nil at 105 dilution	Nil at 105 dilution	Nil at 105 dilution	Nil at 106 dilution
Expiry period from date of manufacture	6 months	6 months	6 months	6months

Parameters	*Rhizobium*	*Azotobacter*	*Azospirillum*	*PSM*
PH	6.5 – 7.5	6.5 - 7.5	6.5 - 7.5	6.5 - 7.5
Strain	Should be checked serologically	Nothing specific A. chroococcum sp. is mentioned	Nothing specific	Nothing specific
Carrier should pass through	150-212 micron IS sieve	106 micron IS sieve	106 micron IS sieve	106 micron IS sieve
Others	Nodulation test + ve result in 50% or more dry matter than control	Minimum amount of N- fixation not less than 5 mg / g of sucrose utilized	Should produce white pellicle in NFB media at 107 dilution	Should show phosphate solubilizing zones of 12 mm in tricalcium phosphate medium at 105 dilution

(ISI, 1979; ISI, 1986)

Economics of Biofertilizers Production

A number of biofertilizer production units have been started recently particularly in the southern states of our country. Nationalized banks have started their Hi-Tech agricultural programme providing loan and motivated the entrepreneurs to start their own production units. The Government of India is also encouraging this low cost technology by providing a subsidy upto Rs.20 lakhs to start a production unit with the capacity of 150 metric tonnes per annum. However, we are all aware that the success of the project entirely depends on the economic viability. With the objective of giving an overall economics of the biofertilizer production and sales, an approximate estimate is prepared.

Total estimate for starting a biofertilizer production unit with the capacity of 150 metric tonnes/annum.

S. No.	Particulars		Amount (Rs.in lakhs)
I.	Expenditure*		
A.	Capital Investment (Fixed cost)		
i.	Building including cost of site (App. 1200 sq. ft.)	:	12.00
ii.	Equipment and apparatus	:	41.00
B.	Operational cost (variable cost)		
i.	Working capital (Raw materials)	:	10.00
ii.	Staff salary	:	2.04
iii.	Labour	:	2.50
iv.	Electricity	:	0.50
v.	Travelling expenses	:	0.50
vi.	Administrative expenses	:	0.50

S. No.	Particulars		Amount (Rs.in lakhs)
vii.	Interest on loan and depreciation	:	0.70
viii.	Miscellaneous expenses	:	0.26
	Total (variable cost)	:	17.00
	Total investment	:	70.00
	Actual initial investment	:	50.00

* The expenditures does not include the marketing expenses

Expenditure details (Rupees in lakhs)

S.No.	Equipment and apparatus	Qty (Nos.)	Amount (Rs.in lakhs)
1.	Fermentor (200 lit. capacity)	4	26.00
2.	Shaker	2	1.50
3.	Laminar air flow chamber	1	0.60
4.	Autoclave	2	0.30
5.	Hot air oven	1	0.10
6.	Incubator	1	0.10
7.	Refrigerator	1	0.30
8.	Microscope	1	0.75
9.	pH meter	1	0.15
10.	Physical balance	1	0.10
11.	Electronic balance	1	0.75
12.	Counter-poise balance	5	0.25
13.	Sealing machine	5	0.25
14.	Work benches	4	0.30
15.	Plastic trays	50	0.25
16.	Trays (Zinc/Aluminium)	10	0.20
17.	Trolley	1	0.10
18.	Automatic packing machine (optional)	1	9.00
	Total		41.00

Working capital

1.	cost of mother culture	:	0.05
2.	Glasswares	:	0.70
3.	Chemicals	:	2.50
4.	Polythene bags	:	3.50
5.	Carrier materials	:	3.00
6.	Miscellaneous items	:	0.25
	Total	:	10.00

Staff salary

Technical staff (1 No.)	:	9000 x 12	1,08,000
Laboratory staff (2 Nos.)	:	4000 x 2 x 12	96,000
Total			2,04,000

II. Production

60% capacity	:	90 MT per year
75% capacity	:	112.5 MT per year
100% capacity	:	150 MT per year

III. Receipts

Cost of 1 kg of biofertilizer (present Govt./University rate)	:	Rs.25/-
Cost of 90 MT (60% capacity)	:	22.500 lakh rupees
112.5 MT (75% capacity)	:	28.125 lakh rupees
135 MT (90% capacity)	:	33.750 lakh rupees
150 MT (100% capacity)		37.500 lakh rupees

IV. Profitability

Year	Production	Receipt (Lakh Rs.)	Expenditure (Lakh Rs.)	Gain (Lakh Rs.)
I	60%	22.500	50.000	-27.500
II	75%	28.125	18.700*	9.425
III	90%	33.750	20.570*	13.180
IV	100%	37.500	22.630*	14.870
	Profit anticipated after 4 years			9.975

*Every year 10% increase in the expenditure is calculated to balance the price escalation

16

Microbial Agents

Insect control through microbial inoculants

Insect pathogens or entomopathogens include viruses, bacteria, fungi and protozoa. These pathogens when conserved and augumented, would keep some of the pests under check within the economic injury levels.

Advantages of using microbial insecticide

Rapid spread

Power of search for host

Persistence, safe

Aesthetically acceptable control to sub economic levels

Predictable control

Virulence

Easy to produce

Low cost

Good storage and

Easy application are the desirable characteristics of entomopathogens for use as microbial insecticide.

Different groups of microorganisms showing the potential for IPM programme.

1. Bacterial pathogens
2. Viral pathogens
3. Fungal pathogens
4. Protozoan pathogens

Bacterial pathogens

Bacteria which affect insects are broadly classified as spore formers and non spore formers. Spore formers include milky disease organisms, *Bacillus popilliae, B. lentimorbus, B. sphaericus, B. thuringiensis*.

Non-spore formers

Serratia, Pseudomonas, Aerobacter and Streptococcus

Among the *Bacillus* species are promising organisms for microbial control.

1. *Bacillus thuringiensis*

Widely exploited species of *Bacillus*. It is environmentally safe, active on insects resistant to chemical insecticide and selective against pest. *B. t* var. *kurstaki* was isolated from *H. armigera* in 1955. The entomocydal activity of *B.t* has been attributed mainly to the parasporal protein crystals, may vary from variety to variety. The crystal proteins are formed during sporulation stage of the bacterium's life cycle.

It is a crystaleceous spore forming organisms. The crystal proteins are called as endotoxis. There are several varieties of *B. t*

B. t. var *thuringiensis*

B. t .var *lentimorpus*

B. t .var *sotto*

B. t .var *canadensis*

Toxins of *B.t.* are of 3 different types

- ❑ α ***exotoxin:*** water soluble, heat labile toxin, identified as lecithin
- ❑ β - ***exotoxin:*** water soluble, heat stable toxin, highly toxic to several flies. Chemically this is adenosine nucleotide analog of ATP. Also called as thuringiensis. Thuringiensis is a specific inhibitor of DNA dependent RNA polymerase and ATP.
- ❑ Delta endotoxin: Most important, called as crystalline toxin.

Symptoms of bacterial infections

1. The body of the insect or larvae becomes soft or dried to a blackscale.
2. Body usually darkened in colour with dark fluids tissues disintegrated with purified odour.
3. Body becomes milky white in colour with body fluids white. (*e.g.* Infection in *B. popilliae*).
4. Body becomes soft and turns to red colour (*Serratia marcesens* infection).
5. In Lepidopteran insects, insects may be found hanging by prolegs in plants or twigs.

Formulations

	Commercial B.t products	Target pest	Commercial products
1.	*B. thuringiensis bv. kurstaki*	Caterpillars	Dipel, Thuricide
2.	*B.thuringiensis bv. israelensis*	Mosquitoes Black flies	Acrobe, Vectobac
3.	*B. thuringiensis bv. tenebrionis*	Colorado Potato beetle	Novodar Trident
4.	*B. thuringiensis bv. aizawi*	Wax – moth Diamond back moth	Florbac

Disadvantages / Limitations

1. Slow in action than chemicals.
2. Method of application and time of application is critical.
3. Environmental stress – Very much influenced by environmental factors like climate, temperature, pH *etc*; which decides the efficiency of the pathogen.
4. Geographic strain variation.
5. A particular strain effective in one place may not be effective in other parts.
6. Host availability
7. Continuous occurrence of host is must otherwise survival of pathogen is a problem.
8. Presence of antimicrobial or antiviral substances.
9. Plants have antimicrobial or antiviral substances, which may interfere with microbial activity
10. Genetic variability
11. May become ineffective on long run.
12. Problem in mass production
13. Most of them need live host
14. Requires technically skilled persons
15. Host specificity

Advantages

1. Host specific due to host specificity, they are active against that particular organism, can be used to control large scale occurrence.
2. Development of resistance is not a problem.
3. Searching capacity is present.
4. Persistent in conducive environment.
5. Non toxic to other organisms.

Fungal pathogens

Entomopathogenic fungi have their ability to decimate insect host populations, particularly during epizootics. Some fungi have very broad host ranges

e.g. Beauveria bassiana

Metarrhizium anisopliae

Verticillium lecanii

Paecilomyces spp.

2. Metarrhizium

Used in the control of oryctus rhinocerous of coconut. The adult beetle causes damage to spathe of adult trees and young plants. Multiplied near FYM area. First it was named as *Entomopthora anisopliae* and latter renamed as *Metarrhizium anisopliae*. The other species is *M. flavoviride*. *M. a* was recorded in 200 insect host belonging to different groups.

Orthoptera

Lepidoptera

Diptera

Hymenoptera, *etc.*

Commonly called Green muscardine fungus. Mass multiplication is in grubs or in laboratory media. When young, the colonies 1.mycelium is white in colour and after 4-5 days when conidia is produced, colour changes to dark green. Toxins are called *Destructins* A, B, C, D, E and desmethyl destruxin B. The toxins play a important role in development of symptoms and death of the organism. LT 50 value for grub is less than adults.

3. Beauveria

First identified in 1912 more than 10 species are identified. Very few are reported to be effective in the pest control.

B. bassiana is a commercially exploited. Toxins are called Beauvericins. Toxic to mosquito larvae and also to potato beetle.

Symptoms

Sluggish movement of larvae, loss of appetite, partial paralysis, reduction in capacity to lay eggs in females, discolouration – are observed initially. Most important stage is the *Sclerotium* formation stage.

Microbial control of diseases

Bacterial and fungal agents are widely used for the control of diseases. Important and potential antagonistic organisms include*Pseudomonas* and *Trichoderma*.

Pseudomonas

Bacteria are the most common type soil microorganisms, provide many benefit to plants and some of them are antagonistic to several pathogenic organisms in the rhizosphere. Among the potential bacterial antagonist associated with plant roots, the fluorescent *Pseudomonas* are quite abundant in the rhizosphere and colonize roots of a wide range of crop plants.

Since these rhizobacteria also promote plant growth, they are called as PGPR.

PGPR mediated Induced Systemic Resistance (IAR) against insects is observed only in few crops. They affect the growth and development of insects at all stages of growth. They are effective against *Helicoverpa*, leaf folder, sheath blight, rice blast and several other fungal diseases.

Method of application

Talc base formulation can be applied through seed, root, foliar and soil.

Mechanism involved in the control

- Stimulate production of biochemical compounds associated with host defense.
- Salicylic acid produced and lipopolysaccharide of *P. fluorescens* act as local and systemic signal molecules in inducing resistance in plants.
- Induce more production of chitinase and peroxidase enzymes, synthesis of phnolics, lignins, PR proteins.
- In case of insect control they may affect the feeding behaviour of the larvae resulting in reduced larval and pupal weight, increased the larval mortality and malformed adult emergence. Chitinase production is also correlated to the control of insects.
- The onset of ISR against pest and disease associated with increased accumulation of defense enzymes *viz.*, PAL, chitinase, PO, PPO, increased amount of phenol and accumulation of lignin. In addition *P. f.* promoted the plant growth and attracted the natural enemies of leaf folder under field conditions.

Trichoderma

It is the widely used biocontrol agent. Among the species of *Trichoderma, T. viride, T. harzianum, T. virens* and *T. hamatum* are used against the management of various diseases of crop plants.

Advantages as biocontrol agent

- High rhizosphere competence
- High competitive saprophytic ability
- Enhance plant growth
- Induce polysaccharide degrading enzymes
- Ease for mass multiplication
- Broad spectrum of action against various pathogens.
- Excellent and reliable control.
- Environmentally safe.

Factors affecting bioefficacy

Abiotic factors influence the bioefficacy of biocontrol agents in soil. pH, temperature, moisture, soil type, components of soil atmosphere, inorganic and organic constituents of soil, quantity of pesticide applied.

Mechanism of disease control

- Competition

- Antibiosis – Trichodermin, dermadin, tricoviridin, sesquiterpene – are the antifungal metabolites.
- Cellulase production
- Lytic enzymes

Proteases – Proteolytic activity against *Sclerotium rolfsii*

- β-1,3 glucanases – degradation of cell wall, results in leakage of protoplasmic contents which can serve as food material for the antagonist Chitinases – hydrolysis of chitin.

Indirect mechanism

Induced resistance –

Plant growth promotion –

17

Microbial Products

Industrial utilization of microorganism

The term **Industrial Microbiology** refers to the use of microorganisms for industrial purposes. Such things as anticoagulants, antidepressants, vasodilators, herbicides, insecticides, plant hormones, enzymes, and vitamins have been isolated from microorganisms or produced in large quantities by genetically engineering the organisms with foreign genes. In commercial industrial plants, microorganisms are widely used to produce numerous organic materials that have far-reaching value and application.

I. Introduction

Microbial processes have been used by man since prehistoric times in the preparation of traditional microbial processes are the production of wine, beer, vinegar, cheese & retting of flax. The discovery of the microbial world led only to a better understanding of the role of microorganism in these processes and subsequently to the improvement and development of new processes based on the use of microorganisms. Today a variety of microorganisms are used to produce a number of substances on a large scale and this area of microbiology has developed into industrial Microbiology or Microbial Biotechnology.

Microbial processes are used on a commercial scale to provide food, beverages, pharmaceuticals, enzymes *etc.*, All these industrial microbiological processes are enhancements of metabolic reactions that microorganisms were already capable of carrying out, with the goal in most cases of simply over producing the product of interest. The team biocatalysis has been used to desirable the reactions carried out by microorganism in Industrial microbiology. It is defined as the use of microorganism to synthesize to convert a substance to a chemically modified form.

II. Industrial Microorganism

The major organisms used in biocatalytic processes are fungi (Yeast and molds) is certain prokaryotes in particular members of the genus streptomyces. Industrial microorganisms are metabolic specialists capable of being manipulated in large scale culture to produce one or more products in high yield.

- The organism must be capable of growth and product formation in large scale culture.
- The organisms should produce spores or some other reproduction cell form so that it can be easily inoculated into large fernentors.
- It grows rapidly and produces the desired product in a relatively short period of time.
- It must also be able to grow in a relatively inexpensive liquid culture medium.(corn step liquor, whey)
- It should not be pathogenic to humans, plants & animals.
- It must be amenable to genetic manipulation

Industrial Products

Microbes are used in the production of alcohols, organic acids, amino acids vitamins, antibiotics, enzymes, single cell protein & hormones. Using different types of fermentation systems.

Methods of culturing

Generally there are two or three methods of culturing microorganism in large scale.

1. Surface culture method - alcohol production, citric acid production
2. Suborned culture method- Most of the Industrial Fermentations.
3. Semis did or solid state culture- Citric acid production.

IV. Production of Alcohol & alcoholic beverages

1. Alcohol production

Production of ethyl alcohol from sugary material is one of the oldest known microbiological processes. Alcohol is an important solvent and raw material used in a variety of Chemical Industries. Although industrial alcohol is produced synthetically from ethylene, production of alcohol by fermentation of cheap sugary materials such as molasses by yeast is an important industry. Most fruit juices undergo a natural fermentation caused by wild yeasts that are present on the fruit.

Selected strains of *Saccharomyces cerevisiae* are employed for production. The alcohol tolerance and sugar tolerance are important criteria used in the selection of yeast strains. The raw material generally used is either crude cane molasses or beet molassess which contain absolute 50 percent fermentable sugar.

The production process involves the dilution of molasses to a suitable sugar concentration (15-16%), addition of small quantity of nitrogen source (Urea, ammonium sulphate or ammonium phosphate), adjustment of PH to about 5.0 addition of an activity growing yeast culture. The fermentation is carried out in big deep tanks of steel or stainless steel and it is allowed to continued for 24-36 hrs at about 25-30°C. Then the cells are allowed to settle. The fermented wash is then distilled and passed through rectifying columns to recover ethyl alcohol. CO_2 is produced in large amount during fermentation. The yield is 50% of the fermentable sugar concentration. Further purification is done by fractional distillation.

In some distillers, yeast is recovered and used as animal feed. Recently cell recycle technique is followed. Basically this involves the reuse of cell mass that is produced during the fermentation. By this 5-10% of substrate used for cell growth has been saved, saving in cost of inoculum, reduction of fermentation time by 24-36 hrs in a batch fermentation.

2. Alcoholic beverages

Whisky, rum, gin, brandy & vodka are all production of alcoholic fermentation of different raw materials such as grains, molasses, potatoes, fruits and these are the distilled products containing a higher consent of alcohol than wine or beer.

a. Wine

Wine is the most important alcoholic beverage, produced by the fermentation of fruits. The grapes are widely used for fermentation. The grapes are crushed by machine and the juice called must is squeezed out. A white wine is made from the juice of red grapes and from which the skins, containing the red colouring matter have been removed. In making red wine, the pomace (skin, seeds & pieces of stem) is left in during fermentation. In addition, red wine has strong flavour, due to the presence of tannis.

Yeast involved in wine fermentation are of two types:

Wild Yeast - Present on grapes.

Cultivated wine yeast - *Saccharomyces ellipsoideus* added to the juice to begin the fermentation.

- Wild yeast are less alcohol tolerant than commercial wine yeast.
- Wild yeast also produce undesirable compounds affecting of the final product.
- Hence in most wineries, wild yeasts are killed in the must by adding sulphur dioxide at level of about 100 ppm.
- *S. ellipsoideus* is resistant to the sulfur dioxide concentration & it is added as starter culture grown on sterilized grape juice.
- Fermentation is carried out is bigger tanks. (200 & to 200,000 l). The fermentation vessel (oak, wheat, stone or glass lined metal) is fitted with a special one way value to the escape the CO_2 produced during fermentation.
- After 3-5 days of fermentation, wine is drawn off for further fermentation in a new tank for one or two weeks.

Next step is called racking. Wine is separate from the sediment, which contains yeast cells & then stored at lower temperature for aging flavour development & further clarification. After filtration, wine is bottled & stored for further aging. Red wine is stored for longer years for aging. During aging process, complex chemical charges occur, including reduction of bitter components, resulting in improvement in flavour & odour or bouquet. Final alcohol content is 6-14% depending on sugar content of grapes, length of fermentation & yeast strain used.

b. Brewing

The manufacturing of alcoholic beverages made form malted, grain is called brewing. Typical malt beverages include beer, ale porter & stout.

Malt: Prepared from germinated barley seeds & it contains natural enzymes that digest the starch in grains & convert it into sugar. Since brewing yeast are unable to digest starch, the malting process is essential for preparation of a fermentable materials from cereal grains. Malt is prepared by first soaking the grains in water & allowing to germinate at 17°C. After germination is over, the grains are dried at 65°C.The powdered malt is either used as such or mixed with other starchy adjuncts. This process is called malting.

Mashing

After making the powdered green malt either alone with other grains (corn, rice, or wheat) is cooked with warm water & allowed to steep at warm temperatures. During the heating period, enzymes from malt causing digestion of starches & liberate sugars which will be fermented by yeast. Projects & amino acids are also liberated into the liquid. This process is called mashing.

Wort

Nutrient rich medium obtained after mashing is called wort, separated by filtration.

Hops

A herb, derived from female flowers of hops plants are now added to the wort. It is a flavouring ingredient, but it has antimicrobial properties which prevent contamination. Then the wort is boiled for several hours & allowed to cool & transferred to fermentation vessel.

In fermentation vessel, it is inoculated with brewers yeast (*S. cerevisiae* or *S. carlsbergensis*& allowed to ferment at low temperature (10-11°C) for 5-10 days. After fermentation, it is clarified by filtration & then pasteurized.

Brewers yeast are two types

Top yeast & bottom yeast

Top yeast: Remain uniformly distributied in the fermentation wort & are carried to the top by CO_2 generated during fermentation. It is used in brewing of ales. *e.g. S.cerevisiae*, fermentation by top yeast occurs at higher temperatures (14-23^{0}C), with short time (5-7 days).

Bottom yeast: Settled at bottom, fermentation is carried out at low temperatures (6-12^{0}C), & for more time (8-14 days). Then top yeast. eg*S.carlsbergensis* used for beer fermentation.

3. Distilled alcoholic Beverages

Distilled alcoholic beverages are made by heating a fermented liquid at high temperature that volatilizes most of the alcohol. The alcohol is then condensed & collected a process called distillation. A product much higher in alcohol content can be obtained by this process.

Distillation of malt brews yield whiskey distilled wine yields brandy distillation of fermented molasses yield rum distillation of fermented grain or potato yield vodka distillation of grain & juniper berries yield grin.

The distillate contains not only alcohol, but also other vollatile products arising from yeast fermentation. Some are desirable & some are undesirable. The undesirable ones are removed during aging process. The fresh distillate is colourless & where as aged product is often brown or yellow. The character of the final product is partly determined by the manner & length of aging.

4. Commodity ethanol production

Production is a biocatalytic process, it is produced from the fermentation of corn starch. This is used as industrial solvent & also for the production of gasohol, a lead free fuel containing 10% ethanol in gaseoline, which is pollution free. *Saccharomyces, Kluveromyces&Candida* are used in commodity ethanol production.

Wine Production

Winemaking, or **Vinification,** is the production of wine, starting with selection of the grapes or other produce and ending with bottling the finished wine. Although most wine is made from grapes, it may also be made from other fruit or non-toxic plant material. Mead is a wine that is made with honey being the primary ingredient after water. Winemaking can be divided into two general categories: still wine production (without carbonation) and sparkling wine production (with carbonation). The science of wine and winemaking is known as oenology (in American English, enology).

PROCESS After the harvest, the grapes are taken into a winery and prepared for primary ferment, at this stage red wine making diverges from white wine making. Red wine is made from the must (pulp) of red or black grapes that undergo fermentation together with the grape skins. White wine is made by fermenting juice which is made by pressing crushed grapes to extract a juice; the skins are removed and play no further role. Occasionally white wine is made from red grapes, this is done by extracting their juice with minimal contact with the grapes' skins. Rosé wines are made from red grapes where the juice is allowed to stay in contact with the dark skins long enough to pick up a pinkish color, but little of the tannins contained in the skins.

To start primary fermentation yeast is added to the must for red wine or juice for white wine. During this fermentation, which often takes between one and two weeks, the yeast converts most of the sugars in the grape juice into ethanol (alcohol) and carbon dioxide. The carbon dioxide is lost to the atmosphere. After the primary fermentation of red grapes the free run wine is pumped off into tanks and the skins are pressed to extract the remaining juice and wine, the press wine blended with the free run wine at the wine makers discretion. The wine is kept warm and the remaining sugars are converted into alcohol and carbon dioxide. The next process in the making of red wine is secondary fermentation. This is a bacterial fermentation which converts malic acid to lactic acid. This process decreases the acid in the wine and softens the taste of the wine. Red wine is sometimes transferred to oak barrels to mature for a period of weeks or months; this practice imparts oak aromas to the wine. The wine must be settled or clarified and adjustments made prior to filtration and bottling. The time from harvest to drinking can vary from a few months for Beaujolais nouveau wines to over twenty years for top wines. However, only about 10% of all red and 5% of white wine will taste better after five years than it will after just one year. Depending on the quality of grape and the target wine style, some of these steps may be combined or omitted to achieve the particular goals of the winemaker. Many wines of comparable quality are produced using similar but distinctly different approaches to their production; quality is dictated by the attributes of the starting material and not necessarily the steps taken during vinification.

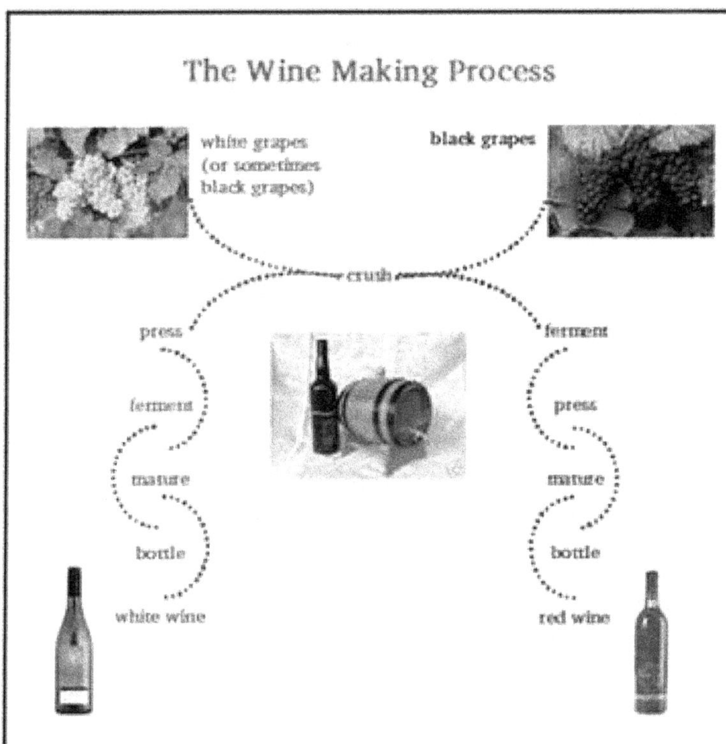

Variations on the above procedure exist. With sparkling wines such as Champagne, an additional fermentation takes place inside the bottle, trapping carbon dioxide and creating the characteristic bubbles. Sweet wines are made by ensuring that some residual sugar remains after fermentation is completed. This can be done by harvesting late (late harvest wine), freezing the grapes to concentrate the sugar (ice wine), or adding a substance to kill the remaining yeast before fermentation is completed; for example, high proof brandy is added when making port wine. In other cases the winemaker may choose to hold back some of the sweet grape juice and add it to the wine after the fermentation is done, a technique known as süssreserve. The process produces wastewater, pomace, and lees that require collection, treatment, and disposal or beneficial use.

Single Cell Protein

Single cell protein (SCP) typically refers to sources of mixed protein extracted from pure or mixed cultures of algae, yeasts, fungi or bacteria (grown on agricultural wastes) used as a substitute for protein-rich foods, in human and animal feeds.

History

Early history Since 2500 BC yeasts have been used in bread and beverage production. In 1781 processes for preparing highly concentrated forms of yeast were established. In 1919 *Endomyces vernalis* yielded fats from sulphite liquor (from paper manufacture), and similarly in 1941 employing Geotrichum. **"Food from oil"** In the 1960s, researchers at British Petroleum developed what they called "proteins-from-oil process": a technology for producing single cell protein by yeast fed by waxy n-paraffins, a product produced by oil refineries. Initial research work was done by Alfred Champagnat at BP's Lavera Oil Refinery in France; a small pilot plant there started operations in March in 1963, and the same construction of the second pilot plant, at Grangemouth Oil Refinery in Britain, was authorized. The term SCP was coined in 1966 by Carol L. Wilson at MIT.

The "food from oil" idea became quite popular by the 1970s, with Champagnat being awarded the UNESCO Science Prize in 1976, and paraffin-fed yeast facilities being built in a number of countries. The primary use of the product was as poultry and cattle feed.

The Soviets were particularly enthusiastic, opening large "BVK" (*belkovo-vitaminny kontsentrat, i.e.,* "protein-vitamin concentrate") plants next to their oil refineries inKstovo (1973) and Kirishi (1974). The Soviet Ministry of Microbiological Industry had eight plants of this kind by 1989, when, pressured by the environmentalist movements, the government decided to close them down, or convert to some other microbiological processes.

SCP Production Process

Single cell proteins develop when microbes ferment waste materials (including wood, straw, cannery and food processing wastes, residues from alcohol production, hydrocarbons, or human and animal excreta). The problem with extracting single cell proteins from the wastes is the dilution and cost. They are found in very low concentrations, usually less than 5%. Engineers have developed ways to increase the concentrations including centrifugation, flotation, precipitation, coagulation and filtration, or the use of semi-permeable membranes. The single cell protein needs to be dehydrated to approximately 10% moisture content and/or acidified to aid in storage and prevent spoilage. The methods to increase the concentrations to adequate levels, and de-watering process require equipment that is expensive and not always suitable for small-scale operations. It is economically prudent to feed the product locally and shortly after it is produced.

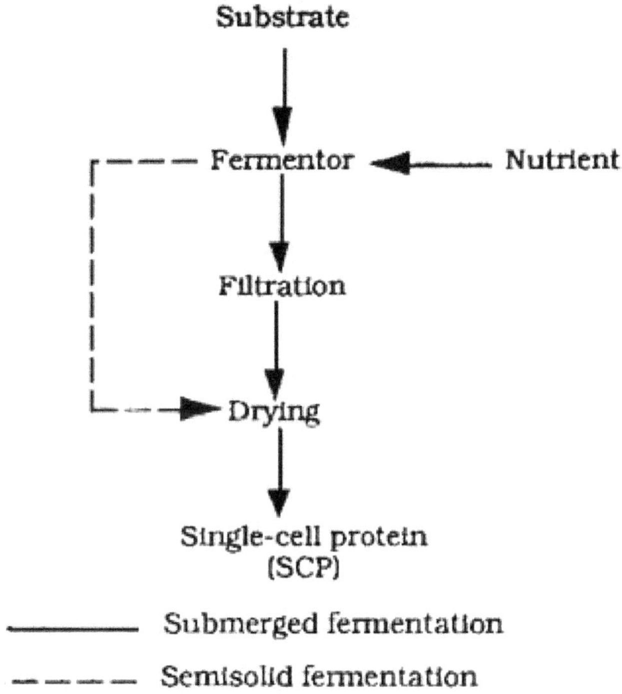

Examples: Microbes employed include yeasts (Saccharomyces cerevisiae, *Candida utilis=Torulopsis* and *Geotrichum candidum*(=Oidium lactis)), other fungi (*Aspergillus oryzae,Sclerotium rolfsii, Polyporus and Trichoderma*), bacteria (*Rhodopseudomonas capsulata*). and algae (*Chlorella* and *Spirulina*). Typical yields of 43 to 56%, with protein contents of 44% to 60%. The fungus Scytalidium acidophilum grows at below pH 1, offering advantages of:

1. low-cost aseptic conditions,
2. avoiding over 100-fold dilution of the acidic hydrolysates to pH values needed for other microbes
3. After the biomass is harvested, the acids can be reused.

Commercial production of SCP (Spirulina) includes Cyanotech in Hawaii and Earthrise in California. TOPRINA- scp made from condidor lipolytica in uk PRUTEEN-SCP made from methanol. TORUTEIN- SCP made from ethanol using torula yeast.

Advantages of Production of SCP

Large-scale production of microbial biomass has many advantages over the traditional methods for producing proteins for food or feed. 1. Microorganisms have a high rate of multiplication to hence rapid succession of generation (algae: 2-6 hours, yeast: 1-3 hours, bacteria: 0.5-2 hours) 2. They can be easily genetically modified for varying the amino acid composition. 3. A very high protein content 43-85 % in the dry mass. 4. They can utilize a broad spectrum of raw materials as carbon sources, which include even waste products. Thus they help in the removal of pollutants also. 5. Strains with high yield and good composition can be selected or produce relatively easily. 6. Microbial biomass production occurs in continuous cultures and the quality is consistent since the growth is independent of seasonal and climatic variations. 7. Land requirements is low and is ecologically beneficial. 8. A high solar energy conversion efficiency per unit area. 9. Solar energy conversion efficiency can be maximized and yield can be enhanced by easy regulation of physical and nutritional factors. 10. Algal culture can be done in space which is normally unused and so there is no need to compete for land.

Cheese

Microbes in Cheese and Yoghurt Milk Protein Consists of proteins, lipids, lactose, minerals, vitamins and Enzymes such as oxidases, phosphatases, peroxidases, catalases, amylases and lipases.Casein makes up 80% of the milk proteincasein is precipitates along with other components when acidified. Milk clotting is done with rennet (chymosin) Rennet hydrolyses the bond between phenylalanine and methionine.

Normal Flora of Cheese Milk

1. Corynebacterium
2. Micrococci
3. Enterococci
4. Spores of Bacillus and Clostridium
5. Staphylococci
6. Coliforms
7. Lactic acid bacteria
   - Lactobacilli
   - Pediococci
   - Leuconostocs

Process of Cheesemaking

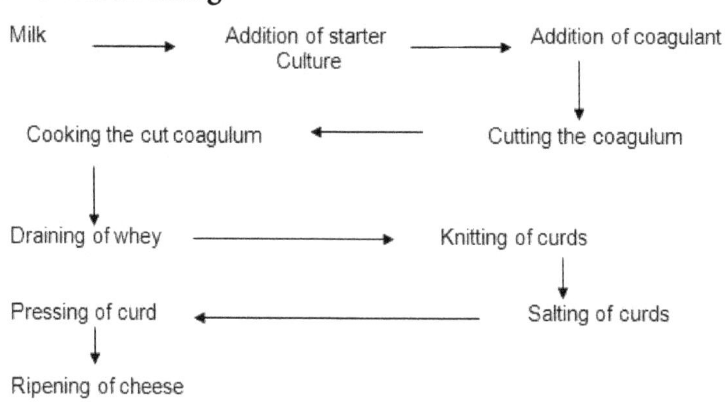

Yogurt

Yoghurt or yogurt is a dairy product produced by bacterial fermentation of milk. Fermentation of lactose produces lactic acid, which acts on milk protein to give yoghurt its texture and its characteristic tang. Dairy yoghurt is produced using a culture of *Lactobacillus delbrueckii* subsp. *bulgaricus* and *Streptococcus salivarius* subsp. *thermophilus* bacteria. The milk is heated to about 80 °C to kill any undesirable bacteria and to change the milk proteins so that they set together rather than form curds. It is then cooled to about 45 °C. The bacteria culture is added, and this temperature is maintained for 4 to 7 hours for fermentation. Soy yoghurt, a non-dairy yoghurt alternative, is made from soy milk.

People have been making and eating yogurt for at least 5,400 years. Today, it is a common food item throughout the world. A nutritious food with unique health benefits, it is rich in protein, calcium, riboflavin, vitamin B6 and vitamin B12.

History

There is evidence of precultured milk products being produced as food for at least 4,500 years. The earliest yoghurts were probably spontaneously fermented by wild bacteria. The oldest writings mentioning yogurt are attributed to Pliny the Elder, who remarked that certain nomadic tribes knew how "to thicken the milk into a substance with an agreeable acidity". The use of yoghurt by medieval Turks is recorded in the books *Diwan Lughat al-Turk* by Mahmud Kashgari and *Kutadgu Bilig* by Yusuf Has Hajibwritten in the 11th century. Both texts mention the word "yoghurt" in different sections and describe its use by nomadic Turks. An early account of a European encounter with yoghurt occurs in French clinical history: Francis I suffered from a severe diarrhoea which no French doctor could cure. His ally Suleiman the Magnificent sent a doctor, who allegedly cured the patient with yoghurt. Being grateful, the French king spread around the information about the food which had cured him.

Raita is a condiment made with yoghurt and popular in India and Pakistan. Until the 1900s, yoghurt was a staple in diets of people in the Russian Empire (and especially Central Asia and the Caucasus), Western Asia, South Eastern Europe/Balkans, Central Europe, and India. Stamen Grigorov (1878–1945), a Bulgarian student of medicine in Geneva, first examined the microflora of the Bulgarian yoghurt. In 1905 he described it as consisting of a spherical and a rod-like lactic acid bacteria. In 1907 the rod-like bacteria was called *Lactobacillus bulgaricus* (now *Lactobacillus delbrueckii subsp. bulgaricus*). The Russian Nobel laureate biologist Ilya Ilyich Mechnikov, from the Institut Pasteur in Paris, was influenced by Grigorov's work and hypothesised that regular consumption of yoghurt was responsible for the unusually long lifespans of Bulgarian peasants. Believing *Lactobacillus* to be essential for good health, Mechnikov worked to popularise yoghurt as a foodstuff throughout Europe. Isaac Carasso industrialized the production of yoghurt. In 1919, Carasso, who was from Ottoman Salonika, started a small yoghurt business in Barcelona and named the business Danone ("little Daniel") after his son. The brand later expanded to the United States under an Americanised version of the name: Dannon. Tarator is a cold, refreshing soup made of yoghurt and cucumber (dill, garlic, walnuts and sunflower oil are sometimes added), popular in Bulgaria. Yoghurt with added fruit jam was patented in 1933 by the Radlická Mlékárna dairy in Prague. It was introduced to the United States in 1947, by Dannon.

Yoghurt was first introduced to the United States by Armenian immigrants Sarkis and Rose Colombosian, who started "Colombo and Sons Creamery" in Andover, Massachusetts in 1929. Colombo Yogurt was originally delivered around New England in a horse-drawn wagon inscribed with the Armenian word "mad zoon" which was later changed to "yogurt", the Turkish name of the product, as Turkish was the lingua franca between immigrants of the various Near Eastern ethnicities who were the main consumers at that time. Yoghurt's popularity in the United States was enhanced in the 1950s and 1960s, when it was presented as a health food. By the late 20th century yoghurt had become a common American food item and Colombo Yogurt was sold in 1993 to General Mills, which discontinued the brand in 2010.

Nutritional value and health benefits

Tzatziki is an appetiser made with yoghurt, popular in Greece and Bulgaria, where it is called Dry Tarator. Yoghurt is nutritionally rich in protein, calcium, riboflavin,vitamin B6 and vitamin B12. It has nutritional benefits beyond those of milk. People who are moderately lactose-intolerant can consume yoghurt without ill effects, because much of the lactose in the milk precursor is converted to lactic acid by the bacterial culture. Yoghurt may also be used in preventing antibiotic-associateddiarrhea. Yoghurt is believed to promote good gum health, possibly because of the effect of lactic acid present in yoghurt. A study published in the *International Journal of Obesity* (11 January 2005) also found that the consumption of low-fat yoghurt can promote weight loss, especially due to the calcium in the yoghurt.

Varieties and presentation

Dadiah sold in Bukittinggi Market Dadiah, or Dadih, is a traditional West Sumatran yoghurt made from water buffalo milk. It is fermented in bamboo tubes. Yoghurt is popular in Nepal, where it is served both as an appetizer and dessert. Locally called *dahi*, it is a part of the Nepali culture, used in local festivals, marriage ceremonies, parties, religious occasions, family gatherings, and so on. The most famous type of Nepalese yoghurt is called *juju dhau*, originating from the city of Bhaktapur. Tarator and Cacık are popular cold soups made from yoghurt, popular during summertime in Albania, Bulgaria, Republic of Macedonia, and Turkey. They are made with ayran, cucumbers, dill, salt, olive oil, and optionally garlic and ground walnuts. Tzatziki, a thick yoghurt-based sauce similar in concoction to tarator, is popular in Greece. Bulgaria typically calls tzatziki "dry tarator".

Khyar w Laban (cucumber and yogurt salad) is a popular dish in Lebanon. Also, a wide variety of local Lebanese dishes are cooked with yogurt like "Kibbi bi Laban" *etc*.. Rahmjoghurt, a creamy yoghurt with much higher fat content (10%) than most yoghurts offered in English-speaking countries (*Rahm* is German for "cream"), is available in Germany and other countries.

Cream-top yoghurt is yoghurt made with unhomogenized milk. A layer of cream rises to the top, forming a rich yoghurt cream. Cream-top yoghurt was first made commercially popular in the United States by Brown Cow of Newfield, New York, bucking the trend toward low- and non-fat yoghurts. Jameed is yoghurt which is salted and dried to preserve it. It is popular in Jordan. Zabadi is the type of yoghurt made in Egypt, usually from the milk of the Egyptian water buffalo. It is particularly associated with Ramadan fasting, as it is thought to prevent thirst during all-day fasting.

Raita is a yoghurt-based South Asian/Indian condiment, used as a side dish. The yoghurt is seasoned with cilantro (coriander), cumin, mint, cayenne pepper, and other herbs and spices. Vegetables such as cucumber and onions are mixed in, and the mixture is served chilled. Raita has a cooling effect on the palate which makes it a good foil for spicy Indian dishes. *Dudh* is a Sindhi-curd, popular in India. People drink dudh along with food at intervals, to help digestion and make food more delicious. In some places dudh is also served with plain rice. *Dahi* is a yoghurt of the Indian subcontinent, known for its characteristic taste and consistency. The word*dahi* seems to be derived from the Sanskrit word *dadhi*, one of the five elixirs, or panchamrita, often used in Hindu ritual. *Dahi* also holds cultural symbolism in many homes in the *Mithilanchal* region of Bihar. It is found in different flavours, two of which are famous: sour yoghurt (*tauk doi*) and sweet yoghurt (*meesti* or *podi doi*). In India, it is often used in cosmetics mixed with turmeric and honey. Sour yoghurt is also used as a hair conditioner by women in many parts of India.

Srikhand, a popular dessert in India, is made from drained yoghurt, saffron, cardamom, nutmeg and sugar and sometimes fruits such as mango or pineapple.

Sweetened and flavored yoghurt

To offset its natural sourness, yoghurt can be sold sweetened, flavored or in containers with fruit or fruit jam on the bottom. If the fruit has been stirred into the yoghurt before purchase, it is commonly referred to as Swiss-style. Most yoghurts in North America] have added pectin, found naturally in fruit, and/or gelatin to artificially create thickness and creaminess at lower cost. This type of adulterated product is also marketed under the name Swiss-style, although it is unrelated to the way yoghurt is eaten in Switzerland. Some yoghurts, often called "cream line," are made with whole milk which has not been homogenized so the cream rises to the top. Fruit jam is used instead of raw fruit pieces in fruit yoghurts to allow storage for weeks. Sweeteners such as cane sugar or sucralose – for low-calorie yogurts – are often present in large amounts in commercial yoghurt. In the USA, sweetened, flavored yoghurt is the most popular type, typically sold in single-serving plastic cups. Typical flavors are vanilla, honey, or fruit such as strawberry, blueberry, blackberry, raspberry, or peach.

Strained yoghurts

Strained yoghurts are types of yoghurt which are strained through a paper or cloth filter, traditionally made of muslin, to remove the whey, giving a much thicker consistency and a distinctive, slightly tangy taste.

Labneh is a strained yoghurt used for sandwiches popular in Arab countries. Olive oil, cucumber slices, olives, and various green herbs may be added. It can be thickened further and rolled into balls, preserved in olive oil, and fermented for a few more weeks. It is sometimes used with onions, meat, and nuts as a stuffing for a variety of pies or kebbeh balls. Some types of strained yoghurts are boiled in open vats first, so that the liquid content is reduced. The popular East Indian dessert, a variation of traditional dahi called mishti dahi, offers a thicker, more custard-like consistency, and is usually sweeter than western yoghurts. Strained yoghurt is also enjoyed in Greece and is the main component of *tzadziki*, a well-known accompaniment to gyros and souvlaki pita sandwiches.

Yogurt products

- Fruit-on-the-bottom style
- Soft-serve and Hard Pack frozen yogurt

18

Antibiotics – Microorganisms Involved & Importance

I. Introduction

Antibiotic production

These are defined as substances produced by some micro-organisms which are in some way lethal to other micro-organisms. It is thought that these substances give the organisms that produce them (usually moulds or actinomycetes - which grow slowly) some sort of advantage in competition with other micro-organisms (usually bacteria - which grow fast) in the same habitat. However, their great medical advantage in healing infections is that the purified forms of antibiotics are more or less harmless to most humans. This means that they must act on some aspect of of the growth of micro-organisms which differs from ordinary mammalian cells. There are in fact several versions of Penicillin, variations on a common formula, produced by different strains of Penicillium, or using different culture media and methods. The **Production of Antibiotics** has been widespread since the pioneering efforts of Florey and Chain in 1938. The importance of antibiotics to medicine has led to much research into their discovery and production.

Identifying Useful Antibiotics

An agar plate streaked with microorganisms. Despite the wide variety of known antibiotics, less than 1% of antimicrobial agents have medical or commercial value. For example, whereas penicillin has a high therapeutic index as it does not generally affect human cells, this is not so for many antibiotics. Other antibiotics simply lack advantage over that already in use, or have no other practical applications.

Useful antibiotics are often discovered using a screening process. To conduct such a screen, isolates of many different microorganisms are cultured and then tested for production of diffusible products that inhibit the growth of test organisms. Most antibiotics identified in such a screen are already known and must therefore be disregarded. The remainder must be tested for their selective toxicities and therapeutic activities, and the best candidates can be examined and possibly modified. A more modern version of this approach is a rational design program. This involves screening directed towards finding new natural products that inhibit a specific target, such as an enzyme only found in the target pathogen, rather than tests to show general inhibition of a culture.

Industrial Production Techniques

Antibiotics are produced industrially by a process of fermentation, where the source microorganism is grown in large containers (100,000–150,000 liters or more) containing a liquid growth medium. Oxygen concentration, temperature, pH and nutrient levels must be optimal, and are closely monitored and adjusted if necessary. As antibiotics are secondary metabolites, the population size must be controlled very carefully to ensure that maximum yield is obtained before the cells die. Once the process is complete, the antibiotic must be extracted and purified to a crystalline product. This is simpler to achieve if the antibiotic is soluble in organic solvent. Otherwise it must first be removed by ion exchange, adsorption or chemical precipitation.

Strains Used for Production

Microorganisms used in fermentation are rarely identical to the wild type. This is because species are often genetically modified to yield the maximum amounts of antibiotics. Mutation is often used, and is encouraged by introducing mutagens such as ultraviolet radiation, x-rays or certain chemicals. Selection and furtherreproduction of the higher yielding strains over many generations can raise yields by 20-fold or more. Another technique used to increase yields is gene amplification, where copies of genes coding for enzymes involved in the antibiotic production can be inserted back into a cell, via vectors such as plasmids. This process must be closely linked with retesting of antibiotic production and effectiveness.

Fermentation

Fermentation is a microbial process by which microbial cell biomass or products made from cells or bioconversion of the substrates can be obtained. The products from cells include primary and secondary metabolites.

Primary metabolites: that is a product excreted during the growth phase of microorganisms. *e.g.* ethanol.

Secondary metabolites: a metabolite which is excreted at the end of the growth phase or in the stationary phase of the cell. *e.g.* Antibiotics, growth hormones.

Antibiotics: are the chemical substance produced by microorganisms; that inhibit or kill the growth of other microorganism even at very low concentrations, it is a product of secondary metabolism. The significance of antibiotic production for the producing strain is unclear. As secondary metabolite, antibiotics could serve regulatory roles during differentiation, perhaps acting as temporary inhibition agents.

Growth inhibition in a mixed culture is known for a long time. *e.g.* Penicillins. Alexander Flemming (1929) was first reported the inhibitory effect of the mold *Penicillium notatum* on the growth of *Staphylococcus aureus*.

Development of antibiotics as agents for treatment of infectious diseases had more impact on the practice of medicine. Large number of known antibiotics are microbial origin and major antibiotics are produced by fermentations and semisynthetic processes. Only very few are produced synthetically.

II. Microbial groups producing antibiotics

Antibiotics are produced by bacteria, actinomycetes and fungi. In fungi, the antibiotics produced by Aspergillaceae and Moniliaceae are of practical importance. Many bacteria are producing antibiotics. Greatest variety of antibiotics are produced by actinomycetes, especially the genus Streptomyces. Two third of antibiotics are produced by actinomycetes only.

III. Classification of antibiotics

Antibiotics are classified according to their

- antimicrobial spectrum
- Mechanism of action
- Producer strain
- Manner of biosynthesis
- Chemical structure

Based on chemical structure antibiotics are grouped as follows

- Carbohydrate containing- Steptomycin, Vancomycin.
- Macrocyclic lactones - Erythromycin , Rifampicin.
- Quinones & related antibiootics - Tetracycline, Adiramycin
- Amino acid and peptide antibiootics- Penicillins, Bacitracin
- Heterocyclic antibiotics containing N- Polyoxins
- Heterocyclic antibiotics containing oxygen- Polyether Antibiotics
- Alicyclic derivatives – cyclohexamide
- Aromatic antibiotics – Chloramphenicol, Griseofulvin.
- Aliphatic anitibiotics – Fosfomycins.

Types of antibiotics based on biosynthesis

Natural antibiotics	-	Normal fermentation yield natural antibiotics.
Biosynthetic antibiotics	-	Production of a particular form of antibiotic by supplying the organisms with specific side chain precursors
Semisynthetic antibiotics	-	produced chemically by adding a specific side chain to the antibiotics
Synthetic antibiotics	-	produced fully by chemical process.

e.g. Penicillin Fermentation

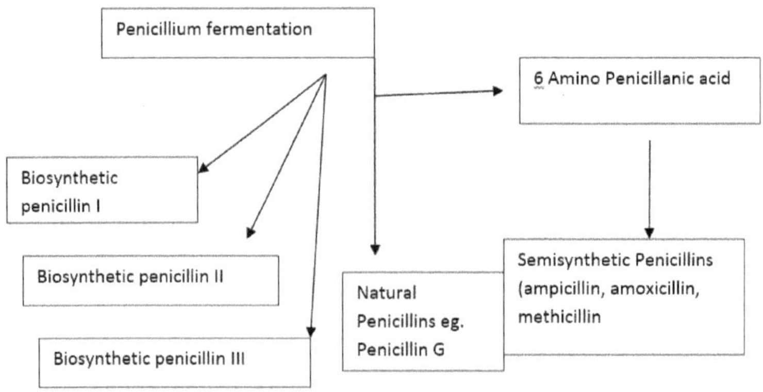

IV. Commercially produced antibiotics

Antibiotic	Microbial source	Antibiotic spectrum	Inhibitory action
Penicillins	Penicillium spp.	Gram- positive bacteria	Cell wall synthesis
Steptomycin	Streptomyces griseus	Gram- positive bacteria and Gram- negative bacteria	Protein synthesis
Tetracycline	Streptomyces aureofaciens	Broad spectrum	Protein synthesis
Bacitracin	Bacillus subtilis	Gram- positive bacteria	Cell wall synthesis
Chloramphenicol	S.Venezuelae	Broad spectrum	Protein synthesis
Cycloheximide	S. griseus	Fungi	Protein synthesis
Nystatin	S.nouresii	Fungi	Cell membrane

V. Penicillin production

It is an important Beta lactum antibiotics and most effective of all therapeutic agents for the control of infectious diseases. It was I described by Flemming in 1929, Florey and Chain at Oxford isolated it from surface cultures of *Penicillium notatum* in 1940 and the I chemical application was in 1941. Natural Penicillins are highly effective against Gram +ve bacteria. They inhibit bacterial cell wall synthesis (peptidoglycon). They combine specifically with Penicillin binding Proteins (PBP) of the bacterial cell, inhibiting the enzyme activity of this protein, bring about cell death.

38% Penicillins produced commercially are used in human medicine, 12% in veterinary medicine and 43% are used as starting materials for the production of semisynthetic penicillins.

Penicillium chrysogenum is used nowadays for the production. Submerged production was followed. Fermentation is anaerobic process lyophilized cultures are used for inoculum development. Initial growth phase is for 40 hrs in prefermenter and it is taken to the production fermenter for Penicillin production. Media contain corn steep liquor (4-5% dry weight) or a commercial N source, soymeal, yeast extract or Whey, and a carbon source lactose is used for production. pH is at 6.5.

Glucose and molasses can also be used for production. After production is over, mycelial mat is separated by filtration and the filtrate is taken to cooling tank and further purification is done. Yield is about 85,000 units/ml.

VI. Applications of Antibiotics

1. Chemotherapeutic use

Chemotherapeutic antibiotics are broad spectrum active against many organisms. Most of them are manufactured as antimicrobial agents.

2. Antitumor antibiotics

They are clinically used as cytostatic agents. eg:

Aclavinomycin	-	Streptomyces galilaeus
Actinomycin D	-	S. antibioticus

3. Antibiotics in plant disease control

Reasons for better use of antibiotics for the control are,

i. applied selectively in low concentration.

ii. Slightly toxic to warm blooded animals and insects.

iii. Easily broken down by soil microorganism.

Table. Antibiotics of use in plant pathology

Antibiotic, chemical type in ( )	Organism producing	Uses
Blasticidin S (Nucleoside)	*S. griseochromogenes*	Rice fungicide against *Piricularia oryzae* (rice burn); relatively toxic
Mildiomycin (Nucleoside)	*Streptoverticillium rimofaciens*	Fungicide for mildew
Polyoxin (Nucleoside)	*S. cacaoi* var. *asoensis*	Multi-purpose fungicide
Prumycin (Nucleoside)	*S. kagawaensis*	Fungicide against *Botrytis* and *Sclerotinia* species
Cycloheximide (Amino acid)	*S. griseus*	Leaf fungicide, highly toxic to warm-blooded animals; aids harvesting
Kasugamycin (Aminoglycoside)	*S. kasugaensis*	Rice fungicide against *Piricularia oryzae*
Validamycin (Aminoglycoside)	*S. hygroscopicus* var. *limoneus*	Fungicide against *Rhizoctonia solani* (leaf dropping and stalk diseases); vegetable crops
Tetranactin (Macrotetrolide)	*S. flaveolus*	Insecticide (mites)

4. Antibiotics as food preservatives

Tylosin	-	effective against *Bacillus* spores
Nisin	-	effective against *Clostridia* are used in Canning Industry.
Chlorotetracycline	-	Used to maintain freshness in fish, meat and poultry by incorporating into ice

5. Antibiotics used as animal growth promoters and in veterinary medicine

Antibiotics are used in sub therapeutic concentrations (1-10 mg/kg feed), weight gain is accelerated, which may causes changes in microflora of the GI tract.

Antibiotics solely used for nutrition purpose

1.	Enduracidin	-	S. fungicidicus
2.	Siomycin	-	S. sioyaensis
3.	Tylosin	-	S. fradiae

Veterinary medicines

Hygromycin B, tylosin, thiostrepton *etc.*

6. Antibiotics as tools in biochemistry and molecular biology

The use of antibiotics as selective inhibitors has made a vital contribution to the understanding of certain cell functions such as DNA replication, transcription, translation and cell wall synthesis. Also used in cell cultures to control contamination. *e.g.* Gentamycin.

18

Biosensor

Biosensor

Biosensor is an analytical device for the detection of an analyte that combines a biological component with a physicochemical detector component.

It consists of 3 parts

The-*sensitive biological element* (biological material (*e.g.* tissue, microorganisms, organelles, cell receptors, enzymes, antibodies, nucleic acids, *etc*), a biologically derived material or biomimic) the sensitive elements can be created by biological engineering.

The-*transducer* or the *detector element* (works in a physicochemical way; optical, piezoelectric, electrochemical, *etc*.) that transforms the signal resulting from the interaction of the analyte with the biological element into another signal (*i.e.*, transducers) that can be more easily measured and quantified;

Associated electronics or signal processors that are primarily responsible for the display of the results in a user-friendly way. This sometimes accounts for the most expensive part of the sensor device, however it is possible to generate a user friendly display that includes transducer and sensitive element (seeHolographic Sensor).

A common example of a commercial biosensor is the blood glucose biosensor, which uses the enzyme glucose oxidase to break blood glucose down. In doing so it first oxidizes glucose and uses two electrons to reduce the FAD (a component of the enzyme) to FADH2. This in turn is oxidized by the electrode (accepting two electrons from the electrode) in a number of steps. The resulting current is a measure of the concentration of glucose. In this case, the electrode is the transducer and the enzyme is the biologically active component.

Recently, arrays of many different detector molecules have been applied in so called electronic nose devices; where the pattern of response from the detectors is used to fingerprint substance. Current commercial electronic noses, however, do not use biological elements.

Principles of Detection Analytical chemistry plays an important role in food quality parameters because almost every sector of industry and public service relies on quality control. A food quality biosensor is a device, which can respond to some property or properties of food and transform the response(s) into a detectable signal, often an electric signal. This signal may provide direct information about the quality factor(s) to be measured or may have a known relation to the quality factor. There are various kinds of biosensors most of which work on the principle of one of the following: **Electrochemical Biosensors** Electrochemical biosensors are based on monitoring electroactive species that are either produced or consumed by the action of the biological components (*e.g.*, enzymes and cells). Transduction of the produced signal can be performed using one of several methods under two broad headings:

- ❑ **Potentiometric Biosensors**
- ❑ **Amperometric Biosensors**

Potentiometric Biosensors These are based on monitoring the potential of a system at a working electrode, with respect to an accurate reference electrode, under conditions of essentially zero current flow. In process, potentiometric measurements are related to the analyte activity (of a target species) in the test sample. Potentiometric biosensors can operate over a wide range (usually several orders of magnitude) of concentrations. The use of potentiometric biosensors for food quality analysis has not been as widely reported as for amperometric sensors. However, some of the examples where this approach has been used for food quality analysis include estimating monophenolase activity in apple juice, determining the concentration of sucrose in soft drinks, measuring isocitrate concentrations in fruit juices, and determining urea levels in milk.

Amperometric Biosensors The use of amperometric biosensors in signal transduction has proved to be the most widely reported using an electrochemical approach. Both- one-shot (disposable) sensors and on-line (multi measurement) devices are commercially available, monitoring a wide range of target analytes. In contrast to potentiometric devices, the principle operation of amperometric biosensors is defined by a constant potential applied between a working and a reference electrode. The applied potential results in redox reactions, causing a net current to flow. The magnitude of this current is proportional to the concentration of electro active species present in test solution and both cathodic (reducing) and anodic (oxidizing) reactions can be monitored amperometrically. Most of the amperometric biosensors described use enzymes as the biorecognition element. Typically, oxidase and dehydrogenase enzymes have been the most frequently exploited catalysts used for these biosensor formats.

Calorimetric Biosensors Most of the biochemical reactions are accompanied by either heat absorption or production. Sensors based on calorimetric transduction are designed to detect heat generated or consumed during a biological reaction; by using sensitive heat detection devices. Various biosensors for specific target analytes have been constructed. In the field of food quality analysis, uses of such biosensors to detect metabolites have been described.

Optical Biosensors These sensors are based on measuring responses to illumination or to light emission. Optical biosensors can employ a number of techniques to detect the presence of a target analyte and are based on well-founded methods including chemiluminescence, fluorescence, light absorbance, phosphoresence, photothermal techniques, surface plasmon resonance (SPR), light polarization and rotation, and total internal reflectance. For example the use of this technique has been demonstrated to detect the presence of allergens, in particular peanuts, during food production.

Acoustic Biosensors Piezoelectric quartz crystals can be affected by a change of mass at the crystal surface; this phenomenon has been successfully exploited and used to develop acoustic biosensors. For practical applications, the surface of the crystal can be modified with recognition elements (*e.g.*, antibodies) that can bind specifically to a target analyte.

Immunosensors Immunosensors are based on exploiting the specific interaction of antibodies with antigens. Typically, immunoassays (such as the enzyme-linked immunosorbent assay technique) employ a label (*e.g.*, enzyme, antibody, fluorescent marker) to detect the immunological reaction. The use of biosensor platforms, linked to an immunoassay format, offers a route to rapid and accurate quantitative measurements of target analytes.

Applications of Biosensors There are many potential applications of biosensors of various types. The main requirements for a biosensor approach to be valuable in terms of research and commercial applications are the identification of a target molecule, availability of a suitable biological recognition element, and the potential for disposable portable detection systems to be preferred to sensitive laboratory-based techniques in some situations. Some examples are given below:

Glucose monitoring in diabetes patients ← historical market driver

Other medical health related targets ◇

Environmental applications *e.g.* the ◇ detection of pesticides and river water contaminants

◇ Remote sensing of airborne bacteria *e.g.* ◇ in counter-bioterrorist activities

◇ Detection of pathogens?

Determining levels of toxic substances ◇ before and after bioremediation

◇ Detection and determining of organophosphate

Routine analytical measurement of folic ◇ acid, biotin, vitamin B_{12} and pantothenic acid as an alternative to microbiological assay

Determination of drug residues in food, ◇ such as antibiotics and growth promoters, particularly meat and honey.

Drug discovery and evaluation of ◇ biological activity of new compounds.

Protein engineering in biosensors?

◇ Detection of toxic metabolites such as mycotoxins

Utility Biosensors for applications in Agriculture in Food/ Fruit Quality Control Quality control is the essential part of a food industry and efficient quality assurance is becoming increasingly important. Consumers expect good quality and healthy food at a given price; with good shelf life and high safety while food inspections require good manufacturing practices, safety, labelling and compliance with the regulations. Further, food producers are increasingly asking for efficient control methods, in particular through on-line or at-line quality sensors. Their main aim is to satisfy the consumer and regulatory requirements and to improve the production feasibility, quality sorting, automation and reduction of production cost and production time subsequently.

Biochemical Composition of Fruits The quality of soft fruit, in terms of taste, nutrition and consumers acceptance, is fundamentally based on the biochemical composition of the fruit. In soft fruits (*viz.* blackcurrant and strawberry) sugar: acid ratios can be used as an important index of fruit maturity and act as a determinant of overall fruit. However, sugar: acid ratios are infrequently used due to a requirement for specific instrumentation and semi-skilled analytical scientists. Today we need a simple and low-cost alternative, which would significantly enhance both the number and extent of tests carried out.

Fruit Maturity, Ripening and Quality Relationships Fruit maturity at harvest is the most important factor that determines shelf life and final fruit quality. If harvested immature then fruits are more subject to shriveling and mechanical damage, and are of inferior quality when ripe, whereas overripe fruits are liable to become soft and mealy with bland flavour soon after harvest. Therefore, fruits harvested either too early or too late in their season are more susceptible to post harvest physiological disorders than fruits harvested at proper maturity.

Fruits can be divided into two groups

Fruit that are incapable of enduring? their ripening process once picked from the plant like berries, cheery, citrus fruits, grapes, lychee, pineapple, pomegranate, and tamarillo.

Fruits that can be harvested mature and ripped off the plant like apple, apricot, avocado, banana, cherimoya, guava, kiwifruit, mango, nectarine, papaya, passion fruit, pear, peach, persimmon, plum, quince, sapodilla, sapota. Volatile compounds are responsible for the characteristic aroma of fruits and are present in extremely small quantities (<100< g/g fresh wt.). The major volatile formed is ethylene. Scientists are trying to develop portable instruments with sensors that detect volatile production by fruits and hence detecting maturity and quality. Other strategies include the removal of a very small amount of fruit tissue and measurement of total sugar or organic acid content.

Major organic acids in fruits Organic acids function in growth, maturation, senescence, color, and antimicrobial activity of fruits. The low pH of fruits is due to the three most common organic acids present in fruits citric acid, malic acid, and tartaric acid. The total amount of acid in fruits varies widely, from about 0.2% in pear juice to 0.8% in limejuice. The amount and type of acid present in fruits determine the fresh taste of fruits and also affects the shelf life.

Organic Acid as an Indicator of Fruit Maturity Organic acids directly play an important role in the growth, maturation and acidity of the fruit, and also affect the shelf life of the fruit by influencing the growth of microorganisms. The citric, malic, oxalic, and tartaric acids ranging from 0.1 to 30 g/L were found in orange, grape, and apple juices. There is a considerable difference in the organic acid content found in various types and brands of fruit juice. For example, Minute Maid contains higher levels of oxalic and citric acids when compared to all other orange juices tested. Grape concentrate was found to have lower amount of malic acid than other grape juice, while freshly squeezed grape juice contains higher amount of tartaric acid. Brae burn apples contained the highest amount of citric acid in apples; however Granny Smith apples were the overall most acidic apples tested.

Successful Examples of Organic Acid Biosensors Developed Pyruvic Acid Onion flavour is principally directed by the perception of pungency. A disposable prototype electrochemical screen-printed (carbon-based) biosensor (C2030519D5, GEM Ltd., Gwent, UK) was constructed using pyruvate dehydrogenase immobilized on mediated Meldolas Blue electrodes and a combined Ag/AgCl reference/counter electrode, both screen-printed onto a PVC substrate to determine pungency in onions (Allium cepa L.). Electrochemical measurements were carried out using a Palm Sense potentiostat (Palm Instruments BV, The Netherlands). The biosensor developed was able to differentiate between mild and pungent bulbs with pyruvate concentrations ranging between 4 and 8 mM in freshly extracted juices. Electrochemical measurements were carried out at +50 mV at 21°C.

Glucose Biosensors

Most of the glucose biosensors developed are based on immobilized glucose oxidase. In many cases, glucose oxidase has been associated with mediators so as to bring down the high working potential required for hydrogen peroxide breakdown. The α-D glucose sensor developed was also based on glucose oxidase, at the working potential of -350 mV vs. Ag/AgCl, hydrogen peroxide was catalytically oxidized at a rhodinised carbon electrode (White et al, 1994). A novel and simple method which do not involve enzyme or monomer modifications, for the coimmobilization of ferrocene and GOx in a poly(pyrrole) matrix for use as glucose biosensor was developed by Foulds and Lowe, 1988. In spite of the low conductivity of the polypyrrole film formed, the biosensor's performance was better than that of other devices reported due to redox mediation of ferrocene that lowers the working potential to 0.4 V. The characterization of the polymer prepared from an ethanolic suspension demonstrated the presence of alcohol interferes in the polymerization kinetics. However, this played a beneficial role in efficient immobilization of both, the enzyme and the ferrocene, in a very thin electroactive film. This fact improved the biosensor's time response, avoiding mass transport effects. A new type of disposable amperometric biosensor was devised by screen-printing thick-film electrodes directly on a porous nitrocellulose (NC) strip. A glucose biosensor based on hydrogen peroxide detection was constructed by immobilizing glucose oxidase (GOx) on the NC electrode strip and by formulating a strong oxidation layer (*i.e.*, PbO2) at the sample loading area, placed below the GOx reaction band. The screen-printed PbO2 paste serves as a sample pretreatment layer that removes interference by its strong oxidizingability. Samples applied were carried chromatographically, via the PbO2 paste, to the GOx layer, and glucose was catalyzed to liberate hydrogen peroxide, which was then detected at the electrode surface. The proposed NC/

PbO2 strip sensor is shown to be virtually insusceptible to interfering species such as acetaminophen and ascorbic and uric acids and to exhibit good performance, in terms of the sensor to sensor reproducibility. The characterization of metal-decorated CNTs was done using X-ray diffraction analysis, transmission electron microscopy (TEM), high-resolution TEM, scanning electron microscopy, and energy-dispersive X-ray analysis. Amperometric biosensor fabricated by depositing GOD over Nafion-solubilized Au-MWNT electrode retained its biocatalytic activity and obtained fast and sensitive glucose quantification. The fabricated GOD/Au- MWNT/Nafion electrode has a good glucose biosensing potential, and it displayed a linear response up to 22 mM glucose and a detection limit of 20 ìM method.

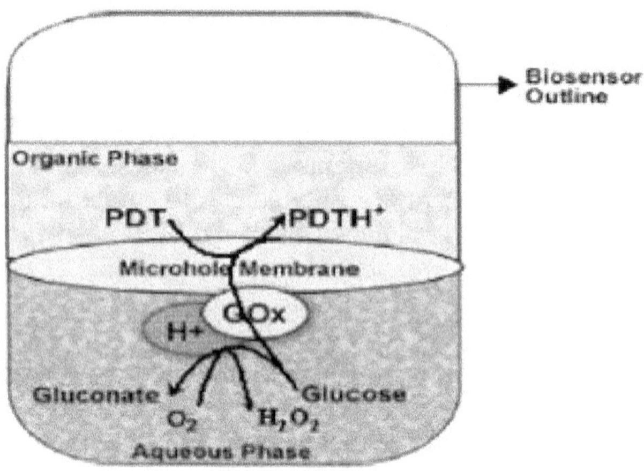

Sucrose Biosensor Sucrose is an essential part of any fruit, so estimating the concentration of sucrose at different maturity levels could help in identifying the ripening parameters of fruits. Therefore, with regard to sucrose detection, electrodes made up of invertase, mutarotase and glucose oxidase and mediated tri-enzyme electrode based on sucrose phosphorylase and electrocatalytic oxidation of NADH, have been used. Because real samples contain both glucose and sucrose, sucrose sensors have been operated in tandem with glucose oxidase sensors. The sucrose sensor developed was based on the invertase, mutarotase and glucose oxidase reaction scheme and the sucrose level was calculated with respect to the net glucose sensors.

Ascorbic Acid Sensor Ascorbic acid has been measured both by direct electrochemical oxidation and by enzymatic methods using ascorbate oxidase. In the first case, an ascorbate oxidase electrode was used to measure the signal generated by other electroactive interferents in the analyte. The second method was based on the measurement of oxygen consumed during the enzyme-catalysed oxidation of ascorbic acid using Clark Electrode.

Lactic Acid Biosensor The level of lactic acid in blood is used in clinical diagnostics of hypoxia, lactic acidosis, some acute heart diseases and drug toxicity tests. Reliable blood lactate measurements would also be of interest in sports medicine. Lactate can be measured based on the reaction using NAD+ dependent lactate dehydrogenase and ferricyanide. The concentration of dissolved L-lactate was determined in tomato paste and baby food samples using a SIRE-based (sensors based on injection of the recognition element) biosensor. The evaluation principle was based on the injection of small amount of enzyme into an internal delivery flow system and held in direct spatial contact with the amperometric transducer by the use of a semipermeable membrane. All the measurements were based upon the reversible enzymatic conversion of L lactate to pyruvate and hydrogen peroxide by lactate oxidase. The L-lactate concentrations of the tomato paste and baby food were calculated to be 1.02 (0.02 mM) and 2.51 (0.10 mM), respectively, using the standard addition method.

Phenolic Compounds Phenolic compounds are widespread in nature, and they play a significant role in living organisms. They are used in medicine and industries, including wood processing and pesticide production. Most of the phenolic derivative compounds are highly toxic, and their determination in low concentrations is the significant problem. Scientists are developing various procedures for determining phenols with biosensors. A biosensor based on crude seed hull enzyme extracts has been prepared for monitoring phenol and hydrogen peroxide. The biosensor has confirmed very promising results as a successful instrument to monitor both hydrogen peroxide and phenol. It is an inexpensive biosensor that could be operated for up to 3 weeks with rapid response and stability parameters. In conditions of response to phenol detection, the developed SBP biosensor was found less sensitive than other previously reported biosensors based on purified SBP or HRP or on crude extracts of sweet potato, which have detection limits in the micromolar range for phenols. The foremost reason for this was the low activity of the enzyme extracts. Further work on the improvement of biosensor sensitivity and applications for the detection of chlorophenols and other substituted phenols are in progress.

The amperometric biosensor described glucose oxidase and polyphenol oxidase carbon paste electrodes prepared via a new strategy of carbon paste modification based on the in situ electropolymerizaton of pyrrole monomer previously mixed within the paste. Such alteration induced a better electrical percolation of the carbon structure and enhanced the enzyme entrapment within the electrode material. Therefore, attractive potentialities offered by a biocomposite electrode based on PPO for the detection of flavonols have been demonstrated to control the phenolic levels in beer samples.

Benzoic Acid Amperometric benzoic acid-sensing inhibitor biosensor was prepared by immobilizing mushroom (*Agaricus bisporus*) tissue homogenate on a Clarktype oxygen electrode. The effects of the quantity of mushroom tissue homogenate, the quantity of gelatin and the effect of the cross-linking agent glutaraldehyde percent on the biosensor were deliberated. The most favourable concentration of phenol used as substrate was 200 mM. The biosensor responded linearly to benzoic acid in a concentration range of 25–100 mM and Standard deviation (s.d.) was found to be ±0.49 µM for 7 successive determinations at a concentration of 75 µM. The inhibitor biosensor based on mushroom tissue homogenate was applied for the determination of benzoic acid in fizzy lemonade, some fruits and groundwater samples. A good concord was shown when the results were compared to those obtained using AOAC method.

Fructose - A superior amperometric biosensor based on a solid binding matrix (SBM) composite transducer has been used for the determination of d-fructose in various food samples. The enzyme, d-fructose dehydrogenase (EC 1.1.99.11), was incorporated directly into a solid composite transducer containing both 2-hexadecanone as SBM and chemically modified graphite. The current variation caused by the presence of d-fructose was calculated amperometrically using Hexacyanoferrate (iii) as a redox mediator. The amperometric signals generated were fast, reproducible and linearly proportional to d-fructose concentrations in the range 50×10-6 – 10×10-**3**mol l-1, with a correlation coefficient of 0.999. A set of measurements at +0.20 V *versus* SCE for 2×10-3 mol l-1 D-fructose yielded a relative standard deviation for the steady-state current of 2.11%. The biosensor selectivity against anionic interferents such as Lascorbate was enhanced by the use of chemically modified graphite by a mild oxidation step. The biosensor was found stable for 6 months and the assay of D-fructose by this electrode was not affected by the presence of sugars or other interferents commonly found in food samples.

Environmental Applications

Toxicity

In environmental pollution monitoring, it is becoming a general opinion that chemical analysis by itself does not provide sufficient information to assess the ecological risk of polluted waters and wastewaters. In the European Union, along with more stringent demands for water treatment (Council Directive 91/271/EEC), industrial and urban wastewater effluents shall reach certain limits of nontoxicity before the effluent can be discharged into the environment. Thus, much effort has been made during the last years to develop and use different bioassays and biosensors for toxicity evaluation of water samples. Whole organisms are used to measure the potential biological impact (toxicity) of a water or soil sample. That is the case of the toxicity assays Microtox® (Azure, Bucks, UK), or ToxAlert® (Merck, Darmstadt, Germany). These systems are based on the use of luminescent bacteria, *Vibrio fischeri*, to measure toxicity from environmental samples.

Bacterial bioluminescence has proved to be a convenient measure of cellular metabolism and, consequently, a reliable sensor for measuring the presence of toxic chemicals in aquatic samples. Some bioassay methods are integrated now in biosensors such as the Cellsense®, which is an amperometric sensor that incorporates *Escherichia coli* bacterial cells for rapid ecotoxicity analysis. It uses ferricyanine, a soluble electron mediator, to divert electrons from the respiratory system of the immobilized bacteria of a suitable carbon electrode. The resulting current is, thus, a measure of bacterial respiratory activity, and the perturbation by pollutants can be detected as a change in the magnitude of the current. Cellsense has been applied to investigate the toxicity of 3,5-dichlorophenol and other phenols in wastewater, for the determination of nonionic surfactants and benzene sulfonate compounds, for the analysis of wastewater treatment works (WWTW) influent and effluent, and for the toxicity testing of wastewaters and sewage sludge.

Moreover, Cellsense has been proposed as one of the newer rapid toxicity assessment methods within the direct toxicity assessment (DTA) demonstration program of the UK Environmental Agency. Most environmental biosensors have focused on bacterial systems while eukariotic biosensors are rare; even more rare is the use of mammalian cells. The mammalian cell, which is more complex than bacteria, can give a more sensitive response when compared to bacteria while also responding to the estrogenic effects of chemicals. A recombinant fluorescent Chinese Hamster Ovary cell line, utilizing a fluorescent reporter system, was used to monitor various toxicants, especially endocrine-disrupting compounds (EDCs), in diverse aqueous environments. EDCs have been also analyzed with a multichannel two-stage mini-bioreactor system using genetically engineered bioluminescent bacteria. The toxicity of various samples spiked with known endocrine-disrupting chemicals, and phenol was investigated.

Conclusions

Despite the huge potential of biosensors, and the ever-increasing number of biosensors developed, commercially available biosensors are being applied to a restricted area of the potential market. In general, biosensors for environmental analysis have several limitations: sensitivity, response time, and lifetime, which should be improved for them to become a competitive analytical tool. The areas of development that are expected to have an impact in biosensor technology are: immobilization techniques, nanotechnology, miniaturization, and multisensor array determinations. However, a crucial aspect may be the production of new sensing elements easy to synthesize and with the capability to broaden the spectra of selectivities that can be reached by a biosensor. At present, the preparation and production in large scales of biomolecules such as enzymes or antibodies need an investment of time and knowledge. Synthetic peptides and MIPs are contemplated as promising alternatives overcoming the above-mentioned limitations.

Unfortunately, the affinity accomplished by these synthetic receptors is still several orders of magnitude below that of the antibodies. Improvement in the affinity, specificity, and mass production of the molecular recognition components may ultimately dictate the success or failure of detection technologies. Possibility of tailor binding molecules with predefined properties, such as selectivity, affinity, and stability, is one of the major aims for biotechnology. The development of advanced receptors will allow the analysis of complex real samples and in situ measurements resolving the responses from the analyte and from nonspecific background effects. Since scientific attention is currently being given to biotechnology, as this review has pointed out, the development of improved molecular recognition elements will be followed by a corresponding enhancement of the biosensor features. From the above viewpoint, it is clear that the future of biosensors will rely on the success of emerging sophisticated micro and nanotechnologies, biochemistry, chemistry, thin-film physics, and electronics.

To reach this goal, an important investment in research, expertise, and the necessary facilities is needed. However, as the world becomes more concerned about the impact that environmental contamination may cause on public health and the ecosystem, the demand for rapid detecting biosensors will only increase.

Biosensors still need to achieve the confidence of potential users, having in mind that the commercialization of new devices will always be the best indicator of the success of a biosensor technology. The analysis of complex matrices and of analytes difficult to determine by the actual analytical procedures (*i.e.*, highly polar compounds), are progressively being approached by biosensors. However, there is still a lack of alternative biosensing systems for an important bunch of emerging contaminants such as bisphenol A, phtalates, and polybrominated compounds (used as flame retardants), veterinary and human medicines and personal care products (nutraceuticals, synthetic fragances, sun screen agents, *etc.*)

Index

www.ingramcontent.com/pod-product-compliance
Ingram Content Group UK Ltd.
Pitfield, Milton Keynes, MK11 3LW, UK
UKHW021323190726
13845UKWH00003B/37

9 789390 371730